How To Write Usable User Documentation

by Edmond H. Weiss

Second Edition

ORYX PRESS 1991

The rare Arabian Oryx is believed to have inspired the myth of the unicorn. This desert antelope became virtually extinct in the early 1960s. At that time several groups of international conservationists arranged to have 9 animals sent to the Phoenix Zoo to be the nucleus of a captive breeding herd. Today the Oryx population is nearly 800, and over 400 have been returned to reserves in the Middle East.

The first edition of this book was titled *How to Write a Usable User Manual*

Copyright © 1991 by
The Oryx Press
4041 North Central at Indian School Road
Phoenix, AZ 85012-3397

Published simultaneously in Canada

Printed and Bound in the United States of America

♾ The paper used in this publication meets the minimum requirements of American National Standard for Information Science—Permanence of Paper for Printed Library Materials, ANSI Z39.48, 1984.

Library of Congress Cataloging-in-Publication Data

Weiss, Edmond H.
  How to write usable user documentation / by Edmond H. Weiss. — 2nd ed.
    p.   cm.
  Includes bibliographical references and index.
  ISBN 0-89774-639-2
  1. Electronic data processing—Authorship. 2. Computers—Handbooks, manuals, etc. 3. Technical writing. I. Title.
QA 76.165.W44   1991
808'.066004—dc20   91-14008
                                                       CIP

For Beverly, Again

Contents

Part 3 Online Documentation and Internal Support 177

Preface

The purpose of this book is to enhance the power and professionalism of everyone who plans, designs, or writes user documentation.

It's hard to believe that when the first edition of this book was published (then titled *How to Write a Usable User Manual*) the IBM PC had just been invented. Today, technical writers and documentors are expected to be "power users" of their computers, facile with word processing, publishing, graphics, communication. They are also expected to be conversant with the human factors literature on such topics as screen design and typography.

The PC/workstation revolution has changed the nature of user documentation. Today, nearly every professional, technical, and clerical worker uses a computer. And they all need *support*. From this perspective, manuals are part of a larger entity: **the user support envelope**. This is an assemblage of publications, Help screens, computer-based tutorials, training materials, interactive videos. . . any information product or service helpful in enhancing the comfort and productivity of users. Typically, today's technical writer is (or should be) responsible for *the whole envelope*.

This second edition has a new title and several additional chapters aimed at *manuals as part of a larger user support context*. These days, manual writing (often called paper documentation) is planned and developed along with online documentation, Help facilities, training programs. . . . Unlike earlier generations of writers, today's writer must even ask whether manuals are necessary. Indeed, we are now entering an era in which the people who *used* to write manuals are often redesigning systems so that they won't need

so much documentation!

The computer technology in use today is of two general types: the old-fashioned, unfriendly kind, which still demands a library of dense, "comprehensive" publications; and the new-fashioned, intuitive kind, which often needs no more than a "minimalist" manual. Each kind wants its own style of support envelope.

This split of product types poses two profound questions for writers.

- First, why are we still laboring to document systems that, with the right menus and Help screens (that is, a better user interface), would need little paper documentation?
- Second, why are we still writing laborious and detailed manuals for well-designed online applications that do not really need them?

Both questions raise issues of power and professionalism. Documentors should be influential members of every system planning team, not just low-level technicians who clean up others' work without asking provocative questions. And documentors should be professionals—people who challenge the approach, schedule, and budgets of the tasks they are assigned.

Writing user documentation is now a *profession*. It is a stimulating mixture of the writer's craft, the artist's design sense, the human factors psychologist's understanding of vision and memory, and the engineer's talent for modeling and testing. The method advanced in this book is a discipline that integrates all these diverse skills.

Acknowledgments

Many people have helped to form the ideas in this book, and many others have supported my research.

I have learned much from my clients and colleagues, including Nurel Beylerian and Mark Ramm of Canada's Institute for Advanced Technology, and Sherry Dell of Digital Equipment Corporation. Much of the first edition of this work was written while I was a consultant to NCR Corporation; my thanks to Stephen Bean, Kenneth Helms, Madeline Flynn, and Michael Bartlett, who not only asked questions but also provided several of the answers.

Thanks also to my son and collaborator, Ryan Weiss, for his assistance with the production of the text. And, finally, deep felt appreciation to Sean Tape, Susan Cain, and Linda Archer of Oryx Press, who tackled the special production problems of this book with enthusiasm and ingenuity and who are responsible for what people in the software business call its "look and feel."

I share the book's merits with all these and others. Its faults are my own.

Edmond H. Weiss
May 1991

Small portions of this book appeared in altered form in Data Training newspaper. These are used with the permission of DATA TRAINING, Warren/Weingarten, Inc.

For permission to reproduce copyrighted excerpts from their user manuals in Appendix D, I thank Brad Solomon, David Champion, Navtel Canada, Falconbridge Ltd., and, again, NCR Corporation.

For permission to reproduce captured screens, I would like to thank the following companies and individuals:

- Ashton-Tate® Corporation, 20101 Hamilton Avenue, P.O. Box 2833, Torrance, CA 90509-9972: Copyright © 1988, Ashton-Tate Corporation. All rights reserved. Reprinted by permission. APPLAUSE II® and SIGN-MASTER® are trademarks of Ashton-Tate Corporation (Exhibits 14.3.1b, 14.5.1b).
- Borland International, 1800 Green Hills Road, Scotts Valley, CA 95066: Quattro Pro screen display courtesy of Borland International (Exhibits 13.3a, 14.3a).
- Brightbill-Roberts & Company, 120 E. Washington St., Suite 421, Syracuse, NY 13202 (Exhibits 14.4.1a, 14.5.3).
- Michael J. Mefford, P.O. Box 351, Gleneden Beach, OR 97388: DirMagic has been upgraded to work with DOS 4 and 5. The upgrade is available to anyone by sending the author a SASE diskette mailer, formatted diskette, and $15 to above address (Exhibit 14.5a).
- Microsoft® Corporation, One Microsoft Way, Redmond, WA 98052-6399: Screen shot(s) Microsoft® Windows™ © 1985–1990 Microsoft Corporation. Reprinted with permission from Microsoft Corporation. Microsoft is a registered trademark and Windows is a trademark of Microsoft Corporation (Exhibits 14.2b, 14.5a, 14.5.2a, and 14.5.2b).
- Peter Norton Computing, 100 Wilshire Blvd 9th Fl, Santa Monica, CA 90401 (Exhibit 14.3.3b).
- Power Up Software Corporation, 2929 Campus Drive, Suite 400, San Mateo, CA 94403 (Exhibits 14.3a, 14.3.4a).
- WordStar® International Incorporated, 201 Alameda del Prado, P.O. Box 6113, Novato, CA 94948: Portions of this book were reproduced with the express written permission of WordStar® International Incorporated (Exhibits 13.3b, 14.3b, and 14.3.3a).

PART 1

Toward a Science of User Documentation

1. TERMS: USERS AND USER DOCUMENTATION

1.1 What Is a User? What Is User Documentation?

*Users are people who must be satisfied. Organizations or individuals buy and develop technology with some goals, in pursuit of particular advantages. Generally, the users are the ones who must be convinced that the goals have been met and the advantages realized. **User documentation** is a collection of information products that help these users get the fullest benefit possible from the technology.*

Any definition of the word *user* is risky. Distinctions between operators and users are fuzzy; even distinctions between users and programmers are getting harder to sustain. With appropriate caution, then, I shall define a **user** as a person more concerned with the *outcome* of information processing than with the *output*. In other words, users are people who treat computers as means to some other ends: business, professional, or personal objectives.

If users are to be satisfied, they must believe that their objectives have been met, with acceptable effort and cost. The end matters more than the means; the outcome more than the output. How the system works is less important than how to work the system to the advantage of the users.

Surely, users can become interested in the inner workings of computers. Indeed, nowadays many users invent their own applications with tools and high-level languages that allow "non-programmers" to succeed at "desktop programming."

But this does not alter the basic idea: Users are people who want something bigger than, and outside of, the particular device they are using. If they could find a cost-effective way to get what they want without a computer, they might.

Why stress this point? Because so many of the people who develop systems—and the associated documentation—tend to view the technology as an end in itself. And because the ensuing user publications and screens are so often unusable technical treatises *about* the product, rather than tools to help the users get what they really want.

For bigger systems and products, the users are often entire organizations, with specialized interests and skills: corporate executives, interested only in the reports; functional managers, seeking administrative support; senior operators, charged with keeping the system going; junior operators and clerks, feeding the system data and monitoring its performance reports; maintenance technicians and programmers; auditors and quality assurance specialists . . .

What all these diverse groups have in common is that they must be *satisfied*. For those who sell computer products in the marketplace, the users are customers. For those who develop systems and applications within their own organizations, the users are the managers of the functional departments. In both cases, the users pay the salaries of the developers.

In the 90s, **user documentation** is a set of information products—manuals, training materials, keyboard templates, online files, and Help screens—that help users (now audiences and readers) get full benefit from the system. Traditionally, user manuals compensate for the difficulty and unfriendliness in systems; they answer such questions as: What do I do next? What does this mean? What is it doing now? Why didn't that work?

In well-planned user documentation, the information products meet the users' changing needs over time. As Exhibit 1.1 shows, user manuals should not only help the users get started but also stay apace of their evolving interests, ultimately reducing the users' dependence on the developers.

Indeed, among the main concerns for today's documentors is the ironic question: How can we improve the system so as to *reduce* the need for user documentation.

Exhibit 1.1: What Users Need from User Documentation

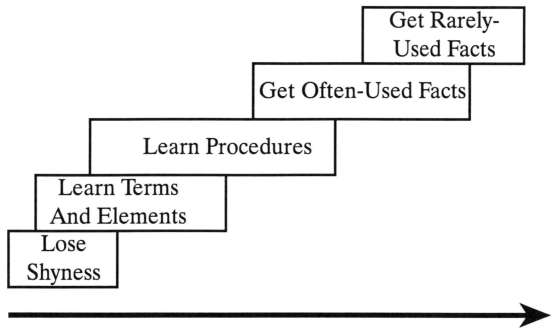

Over Time, Users Need to...

1.2 Why So Many Good People Write Such Bad Documentation

Many of the firms that should write user documentation write none. Most of the firms that write user manuals do not write enough of them or keep them up to date. And many of the manuals written—even by the most sophisticated firms—are ineffective: clumsy, inaccessible, and inaccurate.

There is a growing group of firms that consistently produces high-quality, readable user documentation. A few more firms produce it much of the time. Together, though, they are still a handful.

The more typical case is *no user documentation at all*. Traveling North America, I am still surprised at how many computer companies, engineering firms, software consultants, banks, and manufacturers have no user manuals or operating instructions for their systems or products. (Even more terrifying is how many have *no technical or system documentation* either.)

Those that finally succumb to pressure and try to write documentation are likely to produce unsatisfactory results: books that run the gamut from hastily-typed-and-unusable to expensively-typeset-and-unusable.

Why? How is it that companies are smart enough to design an automated teller or a CAD/CAM system or a network that allows computers to talk to copy machines, but these bright, resourceful organizations cannot manage to write an intelligible user manual?

There are two main explanations: first, some don't care; second, some don't know how.

Long before there were user manuals for computers, there were instruction books and assembly guides for equipment. And for as long as there has been such literature, much of it has been unreadable. Why? Because, traditionally, engineers and manufacturers do not like to spend time or money on these documents, often forcing their publication managers to beg for funds.

Moreover, a good many engineers, scientists, and systems analysts *hate* to write. And the writing they hate the most is explaining complicated, technical ideas to people who know less than they do.

That many firms are indifferent to user documentation is apparent. They set aside almost no time to get it written and often assign it to people with other "more urgent" things to do. Or they delegate it to a junior employee who has never written a complicated publication before and who lacks the authority and leverage to do it well.

Ironically, the writing of online documentation is often entrusted to the same programmers who wrote the cryptic screens and messages that send users to their manuals in desperation.

In those firms that *do* care, matters are a little better. Still, though, the central problem affecting the writers of user manuals—including professional technical writers—is that they have not received enough guidance and instruction on how to write them. Most people about to write a manual have never written one before; only a few have a "good one" to refer to as a model.

Even though there is about 40 years' worth of research on techniques that make documents more accessible and readable, most people, including more than a few professional technical writers, have read none of it. Good writing is still regarded as an art, in the least favorable connotation of the word: a discipline dependent on hunches, intuitions, and instincts. Too many discussions about user manuals—especially about editing and refining them—devolve into disputes about personal preference.

So, in the extreme, stereotypical cases, user manuals are often written by technical experts, who dislike the job, give it as little effort as possible, and use no formal criteria to decide if the job was done well. Or, at the other extreme, they are written by artisan technical writers, who bring all their intuitive and stylistic sense to the project, but who lack the theories and formal criteria needed to evaluate it or to justify its cost to the skeptics.

Analysts and technical experts cannot, working alone, produce usable user documentation. Not because they write badly; they do not write any worse than people in other learned professions. Rather, because they know too much and, with few exceptions, assume so much in their documents that they cannot make themselves clear to less knowledgeable readers.

And neither can most technical communicators, who, in the typical firm, must petition for "input" from the developers, thereby dooming their work to errors and omissions. Indeed, *there is an inherent weakness in any user publication conceived and written entirely by one person.* Such a document is hard to test and nearly unmaintainable.

The **usability** of documentation (that is, how appropriate, accessible, and reliable it is) can be defined and measured. And, furthermore, achieving usability demands participation of both technical *and* communication experts.

1.3 The New Notion of a Document

In the 1990s, our notion of a "document" will be considerably revised. Not only will many documents exist in some form other than paper. Even paper documents will be different; writing and reading will become the creation and manipulation of electronic files and "document databases." In effect, documents will be perpetually revised and enhanced.

The lingering prejudice against documentation is mainly a result of its painful difficulty: almost everyone finds documentation irksome and distracting to write; all but a few find it irksome and unrewarding to read. (A large manual is an easy joke on a television comedy.) But another part of the problem is the inadequacy of the documents themselves. Not only are they often badly written—first drafts by hurried and reluctant authors. They are also often inaccurate and, typically, out of date.

Documentation loses most of its value when it is not current, but, unfortunately, systems usually change faster than documentors can keep up with them. Until recently, documentors could not be blamed. In the paper epoch, a flow diagram was drawn with a pencil and a plastic stencil, a decision table was composed on a typewriter, a glossary of terms was compiled by hand—and none of the resulting documents lent itself to rapid change.

Any form of documentation that resists revision, tends to remain unrevised. And unmaintained documentation falls into disuse and disrepute. If most of the bugs in a program are caused by the latest changes, and if the latest changes are not reflected in the user documentation, what practical good is the documentation?

Many of these problems are obviated by current technology. Today, a prototype screen can be revised and redrawn in a minute. A data dictionary facility *reminds* the user that some new terms have come into use without official definitions. A document can be revised a dozen times in an afternoon, with a clean print of each version and an *automatic highlighting of all the changes from version to version.*

Nowadays a **document**, or a piece of a document, is actually a paper view of a digitally stored entity. And digitally encoded entities are far easier to reproduce, revise, interconnect, and otherwise manipulate than any traditional form of communication.

None of this technology has, of course, altered the basic requirements of user support. Sentences still have to make sense; diagrams must be intelligible; ideas must be logical and coherent. But the change is, nevertheless, qualitative and profound.

Documenters who accept the idea that system-related documents are never finished (just as systems are never finished) can shake off their '50ish ideas about publications and their '60ish ideas about system development. *The documentation of a system is in a nearly continuous state of becoming.* The "manuals," whatever their form, may be revised until one second before shipping or installation.

In the era of document databases, publications will become "virtual"—resident in files and utilities, updated module-by-module, with the same kernel materials appearing in manuals, training materials, and online panels.

Exhibit 1.3: Virtual Documents Reside in Document Databases

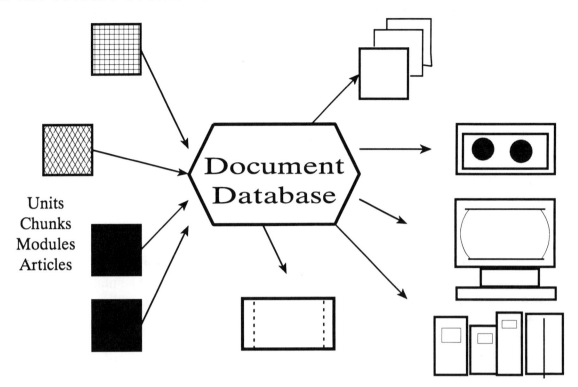

Units
Chunks
Modules
Articles

1.4 An Instrumental Approach: Documents as Devices

If writers are thought of as artists and user documents as works of art, then neither professional writers nor analysts/engineers are likely to produce usable manuals. The key is to think of documents as devices.

If manuals or other information products are thought of as works of art, it will be extremely difficult to change the methods people use to develop them. If, instead, each publication, videotape, or series of Help screens is thought of as a device, with a set of functions, then usability becomes attainable.

Notice the analogy between documents and computer programs. Manuals affect readers the way programs affect computer hardware—except that readers are far more fallible and have far less reliable memories. Manuals or screens pass instructions and data to their readers, who then operate the system correctly and productively.

Thinking of books as devices is a serious change in perspective for many writers and analysts. Doing so obliges them to rethink their notions of user documentation, and to change a whole cluster of related attitudes.

As Exhibit 1.4 shows, the first change required is a new conception of the writing process. If documentation is an art, then the creativity is in the *drafting*, the composing of the words and sentences; writers who think of themselves as artists spend most of their time writing and polishing the draft. In contrast, if a manual is a device, then the creativity is in the *engineering*, writing the specifications and building and testing models—all of which precede the execution of the design (the draft).

A new view of the reader also becomes necessary. The artist views readers as independent and active; the burden is on the reader to find things and apply them correctly. If books are devices, though, readers are less independent. Instead, they rely on the design of the book; the burden shifts to the documentor. In this view the writer controls the attention of the reader, much as software controls hardware—and for similar reasons.

Different criteria should be used for judging publications regarded as devices. If a document is art, then the basic criteria are style and "appeal"—a sense of correctness and craft, peculiarly understandable to the writer but difficult to explain to others. If it is a device, the basic criteria are whether it meets the specifications and performs the job it was assigned.

For artists, a very good book is one that meets the advanced criteria of beauty, elegance, "class." But if a book is a device, the advanced criteria are taken from engineering: maintainability (how easy it is to update and enhance the book) and reliability (how often the book "fails" in use).

And finally, the cost justifications are entirely different. The hardest task for the artist-documentor is to justify the cost of user documentation. Beyond convincing management that at least some user documentation is an unavoidable necessity, the artist is usually powerless to justify expensive processes and products. "Class" and "style" are not usually persuasive. In contrast, the justification for books as devices is that they save or make money: Each device (document) should return more than it costs.

Exhibit 1.4: Documents as Works of Art vs. Devices

	"WORK OF ART"	"DEVICE"
PROCESS	Compose, polish the draft	Spec, test, refine
VIEW OF READER	Independent, resourceful	Dependent, error-prone
BASIC CRITERIA	Style, appeal, preference	Meeting the specs, usability
ADVANCED CRITERIA	Beauty, elegance, "class"	Maintainability, reliability
COST JUSTIFICATION	Unpleasant necessity	Productivity, efficiency, return

2. NEEDS: HOW USER DOCUMENTATION FAILS OR SUCCEEDS

2.1 The Four Functions of User Documentation

Traditionally, user documentation has been divided into two large categories: instruction *and* reference. *Now, user documentation should be divided into four categories—orientation, guidance, motivation, and reference. Thus, instruction becomes* orientation *(tutorials aimed at the novice) as well as* guidance *(demonstrations aimed at the more-experienced user), and a new category is added:* motivation—*writing aimed at overcoming reluctance.*

To say that a manual describes a system or gives information about a procedure is *not* to define its purpose. Very few readers want a "description" or "general information."

Rather, every user publication should perform one or more specific, discernible functions. But what are these functions? What does user documentation do?

The overall purpose of user documentation is to help users get full value from a system—to get their money's worth. Traditionally, user documentation has been expected to help in two ways:

- **instruction**—teaching people how to run or operate the system or product

- **reference**—giving people key definitions, facts, and codes that they could not be expected to memorize

This simple classification scheme worked well during the era in which the typical user was a well-educated engineer, mathematician, or computer professional. Run-books (instruction) and lookup-books (reference) were all that a resourceful user or operator would be likely to need.

But instruction is too large to be considered one category. Instead, I propose to break it into **orientation** and **guidance**. Orientation contains those tutorial materials intended to train neophyte users; guidance includes demonstrations of processes or activities directed to a competent or experienced reader.

Orientation documentation is the newest form of user documentation, and the form that gives the most trouble both to traditional technical writers and, especially, to the programmers and managers who have been conscripted into the job of writing it. Further complicating matters is the rising prominence of a reader I think of as Reader X, a person who is intimidated by books and has seldom been able to learn successfully from reading.

Guidance is teaching by demonstrating and showing. Aimed at a person who knows generally what to do with the system, it shows whole procedures and transactions, from the top down. In contrast, orientation documents ordinarily begin from the bottom, with elemental definitions and concepts.

Reference documentation—what some programmers mistakenly equate with user documentation—is a compressed presentation of facts and information, typically organized alphabetically, useful mainly to people *who know what they need to know.* Highly experienced operators and users need nothing else; new and intermediate operators and users need much more.

The change in the community of users has created the need for a fourth function: **motivation**. Documentation written to provide motivation is supposed to get people to do what they are reluctant to do. In effect, motivation is the selling of ideas and methods. And although not every user manual needs it, far more need it than have it. Put simply, many system problems can be blamed on reluctance, not ignorance. Whether from insecurity or laziness, many operators and users simply will not use systems the way we think they should. They must be "sold."

Exhibit 2.1: Functions of User Documentation

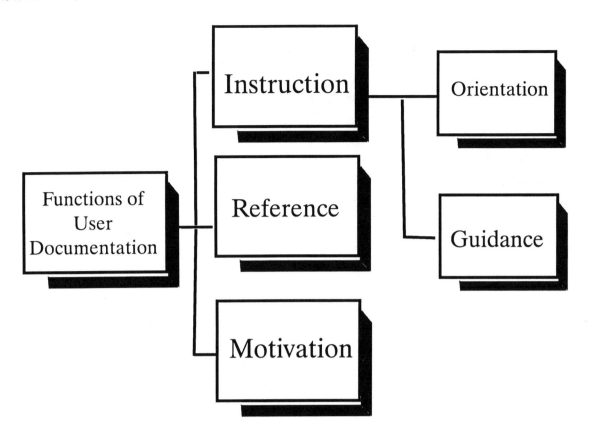

2.2 The Main Goal of User Documentation: *Control*

A paradox of effective writing: To communicate well, one must respect the independence and intelligence of the readers, but must not rely on them. For user documentation, the best strategy is to adapt to the weaknesses in typical readers and to assume control of the communication.

If a user manual is regarded as nothing more than a package of facts, a binderful of miscellaneous information, then its usefulness depends primarily on the skill and resourcefulness of the reader. In contrast, if a manual has been engineered to suit the interests and ability of the reader, then to some extent it *controls* the user, that is, prevents the user from misusing the material.

Many object to this use of the term *control*. Any talk of "controlling people" elicits sincere objections. But this is not to suggest that writers should coerce the behavior of readers. Rather, I mean that anyone who wants to write effective user documentation should regard the readers as complicated information processing systems and try to *control for the sources of noise and error in such systems.*

Documents affect readers in much the way that computer programs affect computers: they *control* their operations. And just as underdesigned software will cause the system to balk or shut down, or to consume too many expensive resources, so will underdesigned documents cause readers to get lost, make errors, even shut down their work. Just as an undertested program will throw off indecipherable bugs nearly every time it is used, so will an undertested manual or menu generate mistakes and inconsistencies.

The objective is control of the readers/ users . . . for their own advantage. The aim is to help the readers gain benefit from the system. And the safest, most reliable way to do that is to devise documents that compel readers to find what they need, in the most efficient sequence,

and with a level of effort that neither discourages them nor lowers their productivity. (I do not recommend this view for all writing, or even all business writing. Literature depends on the imagination, experience, and intellect of the reader, often demanding close reading and study. But user manuals that must be studied to be understood are, in general, ineffective.)

Every user publication fits somewhere on the continuum that appears in Exhibit 2.2. At the highest level of control are those publications meant to be read from the first word to the last, without omissions, without skipping or skimming. Most notable in this group are installation plans, assembly instructions, orientation materials, new product proposals, and specifications. In this category is nearly every document that is **incremental** (presenting an accumulation of increasingly complicated facts), **procedural** (presenting a set of steps or activities that constrain each other), or **argumentative** (presenting a logical chain of assertions).

At the other extreme are publications that no one would ever read in sequence: dictionaries, glossaries, inventories, and directories—alphabetical or numerical listings of reference material. Yet, even at this end of the continuum, there is still a benefit in controlling the reader. A well-designed reference directory allows the user to find information quickly, with "one pass," to complete the search without needing to skip and detour, and, finally, to exit promptly with the needed information. Underdesigned documents increase the **document overhead**: the ratio of the effort needed to *find* information to the effort needed to *use* it.

Exhibit 2.2: Continuum of Control

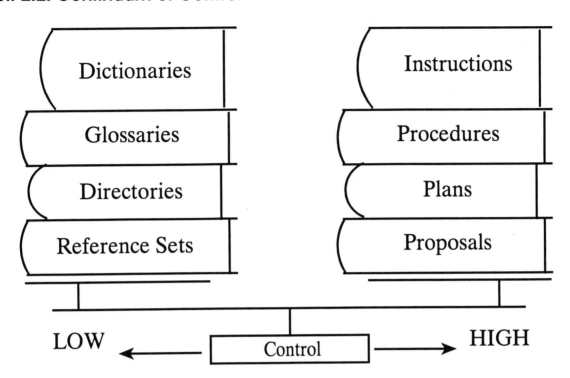

2.3 Four Criteria for Effective User Documents

Documents (or designs for documents) can be tested against formal criteria. If the organization can agree on the criteria, it can then develop quality metrics. The most useful criteria for judging publications, from the least to the most demanding, are availability, suitability, accessibility, and readability.

There are at least four levels of documentation quality, starting with **availability** (Is there anything at all?) and moving through **readability** (Is it in clear, easy-to-understand English?).

Availability

There are still developers who provide no user documentation (or nearly none). Typically, these are organizations in which almost everyone is a programmer. Such organizations simply are not attuned to users—what they do, what they know, how they work. And until the unit hires someone with such an awareness, it will continue to overlook user documentation.

Suitability

Today, most developers provide at least some user documents. Sadly, though, they tend to subscribe to the encyclopedic view of the user manual: Put everything in one big volume and let

Exhibit 2.3: Quality Criteria for User Documentation

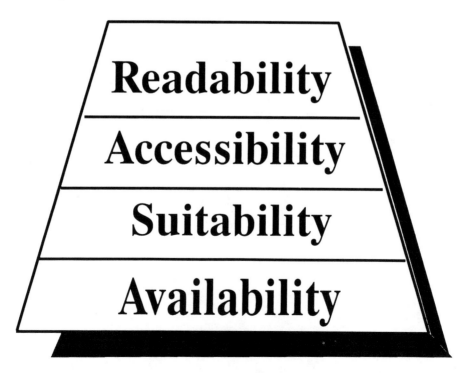

| Readability |
| Accessibility |
| Suitability |
| Availability |

the users fend for themselves. What they ought to do is *analyze* what documents are needed and align particular publications with the tasks and interests of particular readers. Until that happens, their "comprehensive" documents will often be unsuitable, unusable, and unreliable.

Accessibility

It is possible for a book to contain exactly what the user needs, but still to be organized in a useless tangle. As a result, readers have to skip, branch, loop, and detour from page to page—until they get lost. In software engineering terms, the book, because of its excessive number of GOTOs, is unreliable. Even a skillful reader will probably get lost.

Only firms that *design* their books for accessibility (and test and debug the designs) produce smooth-reading, GOTO-less user manuals. A user manual that is both suitable and accessible is likely to be called **task-oriented**. This means that the developer of the manual has analyzed what the users do, how they use the system and product, and what information they need.

Interestingly, the '80s began with a cry for "comprehensive" documentation and ended with a fascination for "minimalist" documents.

Readability

Even when a book is suitable and accessible, its ultimate quality resides in its readability— how easily and accurately it can be understood by its intended audience of users. Still, too many regard matters of language and style as "frills"; hundreds of manuals and instruction books are published without so much as a cursory review by a professional writer or editor. Only *professional editing* can produce manuals of the highest quality.

Note: Although every sentence in every publication should be as readable as possible, well-written sentences offer no real benefit to usability if they are the wrong sentences or are in an unworkable arrangement. *You cannot make old manuals usable merely by improving their style.*

2.4 Three Classes of Error

For user documents to score high on the four criteria of quality, they must be well designed. When documents score low on one or more of the criteria, the failure can be blamed on one or more classes or errors: strategic, structural, *or* tactical.

A point rarely appreciated is that *much of what is wrong with user documentation is the result of mistakes made before the draft was written.* It is equally true that *the most serious flaws in user publications are nearly impossible to correct after the first complete version of the publication is drafted.* Astonishingly, in some firms, editors don't even see the manual until the programmers and analysts consider it finished!

There are three broad classes of error that can undermine user documentation—and only the last of them can be corrected in the editing stage. The first, **strategic errors**, includes failures of planning and analysis: failure to define what documents were needed to serve the likely audiences in the completion of specific sets of tasks and applications. The main strategic errors are

- overlooking the need to plan or analyze documentation requirements

- allowing the product or system to shape the documentation, instead of the users' interests and tasks

- assuming that *only one encyclopedic manual is needed*

- refusing to adapt to the vocabulary and reading skills of the intended audience

Structural errors are failures of design and modeling: insufficient outlining, lack of rigorous review of the outlines, failure to test the plan of the publication before writing a detailed draft. Even if the planners have made no strategic errors, structural errors can still lower the suita-

bility of the manuals and, more relevant, so reduce their accessibility as to make them unusable. The most common structural errors are

- using little or no outlining or other document specifications

- relying on superficial, "grade school" outlining methods

- failing to submit outlines and specifications to harsh reviews (walkthroughs)

- excluding the intended users and readers from the design process

Tactical errors are failures of editing and revision: inconsistent nomenclature, mechanical errors of grammar and spelling, clumsy "first draft" style, ambiguous sentences. Tactical errors occur either when the organization lacks competent editors or when it just does not allow enough time for the editors to work.

Notice the paradox. On the one hand, it is a serious mistake to publish a manual that has never been reviewed by a competent wordsmith. On the other hand, is even more dangerous to believe that the skills of a wordsmith can compensate for having written the *wrong* publication.

In effect, then, a usable manual must pass three tests:

- The *strategic test* proves that the manual is well-defined, is aligned with a specific audience and use, and is part of a coherent set or list of information products.

- The *structural test* ensures that the elements in the publication are in the most accessible, reliable sequence.

- The *tactical test* ensures that the sentences and diagrams will be free from distracting errors and clumsy style.

Exhibit 2.4: Three Classes of Document Error

ERROR	SOURCE
STRATEGIC Failures of Planning/Analysis	▪ poor definition of audiences ▪ poor definition of tasks ▪ lack of overall support plan
STRUCTURAL Failures of Design/Modeling	▪ lack of substantive outlines ▪ poor tests of outlines ▪ excluding users from the review
TACTICAL Failures of Editing/Revision	▪ careless inconsistencies ▪ "first draft" style ▪ substandard editing

3. USABILITY: DOCUMENTATION AS A SYSTEM

3.1 From "Idiot-Proof" to "Usable"

When engineers and inventors devise truly new products or techniques, they frequently worry least about whether the product is easy to use. Ever since the mid-1980s, though, usability *has become one of the main objectives for designers of computer and communication products.*

Today's computer systems for the most part perform the same tasks as the computers of the 1950s—but do them faster, cheaper, and with less human effort. Since the advent of the data-processing industry, then, there has been an evolution of criteria; with each era, the ante has been raised.

As Exhibit 3.1 shows, the prevailing criterion of system quality in the 1950s was mere **performance**—whether the system worked at all. Gradually, analysts and engineers shifted their attention to the economies of **efficiency**—throughput and cycle times, resources used, and so forth.

As machines and memory dropped in price, though, the emphasis on efficiency decreased in many places. Nowadays, it often costs more in personnel expenses to make a machine efficient than could be saved in the efficiencies. Today,

the most important, most frequently discussed technical criterion is **maintainability**, the ease with which a system can be fixed, adjusted, or enhanced.

In the 1980s, the theme changed somewhat. Although many organizations had still not entered the 1970s—that is, they were still concocting unmaintainable systems without benefit of the new development methods—the latest chant was "user friendliness." The criterion became **usability**—making the system easy to use.

Computer technology, then, has completed an entire cycle of development: It still does mostly the same things it did in the beginning—but in a much friendlier manner. The typical operator of today's computer is not a mathematician or programmer, but rather a clerk or business person, or even a 10-year-old child. Engineers no longer use such terms as "idiot-proof" to describe

Exhibit 3.1: Evolution of System Quality Criteria

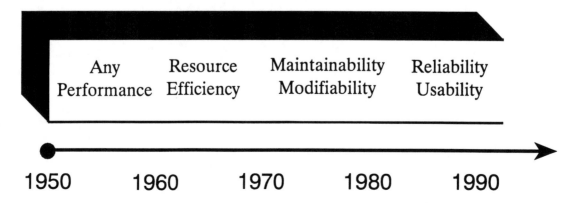

new systems, for to do so harks back to an earlier epoch of computer technology, when the user was presumed to be an expert.

Usability is an engineered constraint. That is, the built-in characteristics of a device, system, or program put an upper limit on how easy that entity will be to use. A task that calls for 20 keystrokes usually will be more error-prone than a task that calls for 2—no matter how well the instructions are written. A <Clear Display> key right next to an <Insert> key is more likely to produce an inadvertent clearing of the display than a key several millimeters removed, despite a warning in the manuals.

There are competing notions of usability, of course. For example, making a system easier to learn at first is not always consistent with making it easy in long-term everyday use.

Usability is a consequence of how well the system has been defined, specified, and tested. It comes from doing the analysis and design well, *not from writing heroic user documentation after the fact.*

For documentors, moreover, the term usability has two important, related meanings. First, it refers to the ease with which a system can be operated; second, it refers to the ease with which the documentation can be operated. Put another way, if the user documentation is also regarded as a system of communication devices, then it follows that the *usability of the documents restricts the usability of the computer system.* When the user documentation is extremely usable, then the computer system will be no harder to use than it must be. If the set of user manuals and other information products is the best possible, of high usability, then the system documented will be as easy to use as its engineering permits.

3.2 The First Law of User Documentation

Each system or product has an inherent usability; each document associated with the system has its own usability. But even the best documents cannot compensate effectively for flaws in the system itself. The first law is **Clean Documentation Cannot Improve Messy Systems.**

Once a system is installed, there is little anyone can do to change its overall usability. Although it can be improved, the improvements are likely to be superficial. One especially ineffective way to increase the usability of an existing system is to try cover its flaws with "especially good" user documentation (in effect, leave the hole in the road but post warning signs, and then mistakenly believe that the hole is no longer a danger).

Drawing a map of a jungle will not turn it into a garden. Nor will writing a slick operations guide make an intimidating and complicated procedure usable. Whenever user documentation is planned after-the-fact (first you develop, then you document), it cannot compensate for failures of analysis, design, or coding.

Although it may be odd to say so in a book about user documentation, it is wrong to expect user manuals to do too much. They should not be expected to ameliorate engineering and programming mistakes.

Clean documentation cannot improve messy systems. Please remember that a simple procedure, explained well, is clearly simple. A difficult procedure, explained well, is still difficult. A dangerous and trouble-prone procedure, explained well, is clearly dangerous and trouble-prone. Just because bad writing makes procedures harder to follow, it does *not* then follow that good writing will make them easier to follow.

The best way for user documentation to improve a system is for it to be created **integrally with the system,** that is, for a "user support envelope" of information products and services to be planned as part of the system itself. Then writers of user documents, as the "first users" of the system, can discover ways to improve the system that developers are unlikely to see. And if they write clearly enough, before the system is etched in disk, there may be time to modify the system.

As Exhibit 3.2 shows, if user documentation is written (or at least designed) during the functional specification of the system, it can be used as an engineering tool; developers can detect and correct errors and unreliabilities in the human part of the system—the so-called user interface. Even during the design stage, there is still a chance that the discovery of hard-to-explain procedures can be reflected in improvements within the modules of the system being documented; it is still practical to make these changes. At the trailing end of development, however, the documentor is more or less stuck with the system as it is.

Note the irony: Documentors who discover flaws in the systems soon enough can eliminate many pages of tortuous documentation.

Exhibit 3.2: Effects of Documentation Phasing

IF DEVELOPED DURING...	DOCUMENTATION CAN:
Functional Specification	■ Clarify procedures and policies ■ Identify unreliable elements ■ Increase chances for user satisfaction
Product Design/Coding	■ Expose bugs and errors ■ Suggest more efficient designs ■ Get designers to make early decisions
Distribution and Use	■ Help users adapt and accept ■ Warn against bugs in the system ■ Disclaim liability

3.3 Defining and Measuring the Usability of Publications

If the objective is to design and engineer publications for usability, and if the process is to be more than "artistic," then there must be formal testing—not only in the finished state but also at intermediate stages. A proposed Index of Usability: **The more often the intended reader must skip material or reverse directions while reading, the less usable the publication.**

Although the most usable publication in the world cannot compensate for inadequacies in a system, usable documentation is an essential ingredient in successful implementation. Is it possible to define a formal Index of Usability for documentation, in such a way that it can be applied *before manuals are written*, in time to correct whatever flaws and bugs it discloses? (Remember the essential point: The later in the life of an information product, the more expensive changing it becomes and, therefore, the less likely that it will be changed.)

I propose that the most predictive Index of Usability is *the number of times the intended reader must skip material or reverse directions to use the publication*. Obviously, this is an *inverse* predictor: the more skipping and looping, the less usable the publication. Of the two, reversing directions (looping) is the far more serious flaw. Reading is both continuous and one-directional; anything in a document that either breaks the continuity or reverses the normal direction reduces the efficacy of the reading process and makes the book less reliable.

This Usability Index is not meant to suggest that all user documentation should be written so that every user reads straight through, from the first word to the last. In fact, relatively few publications will be like that. The point, rather, is that any skip or loop in the document—intended or not—exacts a cost and lowers usability.

Note also that the proposed Usability Index includes the phrase "intended reader." Clearly, readers with different interests and backgrounds would use the same publication differently. Indeed, the same reader, after one or two one-directional passes through a manual, would later skip and glean. Clearly, the more diverse the audience for a certain manual, the harder it is to make it usable for everyone.

Interestingly, the skips and loops (branches, detours, and GOTOs) can be grouped into three classes, corresponding to the three main errors of documentation:

Strategic errors (errors of boundary and scope) cause the largest skips and spins. Failure to align the books with the readers will send the readers jumping from book to book, until they finally find what they need—or give up. If a user needs two books to do one job, the selection and partitioning of the books does not reflect the needs and interests of *that user*. And if a user must ignore 98 percent of a publication, it must have been designed for someone else.

Structural errors cause medium-sized loops and skips. Even though the publication has the right content, it calls for frequent jumping from front to back, especially when the text refers to charts, tables, and exhibits that are elsewhere. *Among the greatest barriers to the usability of a publication is the separation of the text from the exhibits* *referred to in the text.* Readers should be able to *see* Exhibit 1 when the text says "See Exhibit 1."

Tactical errors cause the smallest GOTOs, usually within a paragraph or page. Because the editing is poor, the reader must loop on unclear sentences, inconsistent nomenclature, distracting errors of grammar, and so forth. Although these are the smallest breaches of usability, they can be powerful enough to undermine even the right book with the right structure. (Defined in this way, only tactical errors need await completion of the draft before detection.)

Exhibit 3.3: Error Types and Their Associated Loops

ERROR-TYPE	LOOP-TYPE
Strategic	■ searching several books ■ needing two books for one task ■ needing to ignore most pages
Structural	■ jumping from front to back ■ never reading pages in sequence ■ searching for exhibits, tables…
Tactical	■ stopping to notice mechanical errors ■ getting stuck on inconsistent terminology ■ rereading difficult passages

3.4 Usable Manuals Are Task-Oriented

Product-oriented manuals are usually horizontal; they describe everything that could be done and are usually organized according to the characteristics of the product described. Task-oriented manuals are vertical; they show how to do specific things and are organized according to the procedures or tasks to be carried out by the reader.

In the early days of computing, languages did everything, while application programs did one or two things; there were few user-defined options. So the run-books for these programs were straightforward, linear, easy to follow, and task-oriented.

Today's products, however, perform as many applications as you can think of: high-level languages; database management systems; packages that do nearly any statistical analysis or generate nearly any common business chart; "front-end" packages that connect the packages.

There is a potentially significant strategic problem in writing the user documentation for these multipurpose systems. Many users are unresponsive to the discussions of dozens of generic skills and features. For example, a doctor or warehouse manager may be uninterested in "How to Write a Column Formula," but exceedingly interested in how to define a particular column in a particular spreadsheet.

Here is the paradox: Even the most versatile software products—whether they are database managers or spreadsheets, or "integrated multi-tasking programming environments"—are used in *applications*. Although the people or company who invented the product may be terribly proud of its versatility, and even though some sophisticated users (mainly experienced computing

hands) can think of a hundred useful things to do with the product, most users want to learn how to do *their* projects, solve *their* problems, and improve *their* performance.

Horizontal (product-oriented) publications usually reveal themselves in their tables of contents. The document is arranged alphabetically (by program, command, transaction, or feature), or sometimes by a logical grouping of parts and components (for example, front panel, back panel, keyboard, buffers) To find information in these documents, users must *know what they need to know.*

In contrast, vertical (task-oriented) manuals have tables of contents with language and operations familiar to the readers. If the users know what they *have to do*, the publication tells them what they need to know about the system.

Consider the pair of outlines in Exhibit 3.4. Do you notice that they cover many of the same topics? Although there are some interesting differences in style (to be discussed later), the main difference is that Version A is horizontal and Version B is vertical. To use Version A, one must skip and loop incessantly, and this lack of usability in the manual will detract from the usability of the product. Version B is task-oriented. In use, Version B will be much more reliable.

Exhibit 3.4: Horizontal vs. Vertical Organization

Version A (Horizontal)	Version B (Vertical)
1. System Administration	1. Installing Your System
1.1 Defaulting Security Features	1.1 Backing-Up the Distribution Disks
1.2 Defining Configuration	1.2 Defining Your Company's Security Rules
1.3 Initializing Files	2. Creating Your Files
2. File Management	2.1 Setting-Up Your Chart of Accounts
2.1 Defining a File	2.2 Transferring Your Current Books
2.2 Reading Files	2.3 Choosing the Budget "Planning Factors"
2.3 Linking Files	3. Applications
2.4 Updating/Maintaining Files	3.1 Analyzing Profit and Loss by Cost Center
3. Input Preparation	3.2 Analyzing Year-to-Year Differentials
3.1 Worksheets	3.3 Forecasting Revenues and Costs
3.2 Data Entry	3.4 Simulating Alternative Budgets
3.3 Data Editing	3.5 Simulating Return-On-Investment
4. Outputs	4. Presentations
4.1 Printing	4.1 Making TREND Charts
4.2 Graphics Printing/Plotting	4.2 Making SHARE Charts
4.3 Storage	4.3 Making COMPARE Charts
Appendix I Alternative Configurations	4.4 Making WORD Tables
Appendix II Sample Outputs	
Appendix III Error Messages	

3.5 Controversy: Usability versus Economy

Many of the things that make a document more usable lead to repetition and duplication, even what some would call waste. Often, the objectives of usability and economy are in conflict, and the conflict must be resolved through policy and negotiation.

Many of the people who lead the user documentation activity in the their companies are publications managers, whose top priority is often **production economy**, keeping the above-the-line costs for documentation as low as possible.

Sometimes, though, the downstream costs and diseconomies associated with mediocre user documentation overshadow the short-term savings for paper, printing, and mailing. Production economy, more often than not, is in conflict with usability. Many of the practices that reduce the production, distribution, and storage costs of manuals also reduce their readability. For example: Have you ever met a reader who enjoyed working from microfilm or reading fine print? Do you know anyone who likes to flip back and forth between two sections of a book, taking comfort in the fact that the publisher was able to avoid duplication?

Consider first the methods for reducing the bulk of manuals. The combination of small print and narrow margins is the easiest way to reduce the number of pages—and the associated printing, mailing, and filing costs. The result is densely packed, nearly unreadable documents.

Similarly, many editors and publishers object to blank spaces and half-empty pages. But this book advocates (and practices in its own format) the policy of beginning each new section at the beginning of the next page; the result is loosely packed, more-readable documents.

The most controversial issue is redundancy— a term that most of us have seen as a criticism of our reports and essays in school. Yet, redundancy is not always a term of criticism. In engineering,

redundancy refers to the existence of deliberate backups, technology that allows the system to keep working even when the primary device fails or malfunctions. From the extra buttons sewn into a good suit to the three or four extra sources of power to drive the coolant pumps in a nuclear power generator, the idea is the same. Redundancy means reliability.

In communication, redundancy compensates for the noise and entropy in a channel. The safest way to get an undistorted signal through a noisy channel is to send it more than once. That's why pilots repeat themselves when they talk to the flight controllers ("yes, affirmative") and that's why electronic funds transfers are sent at least twice, and then checked for parity.

In some ways, even large typefaces, wide margins, and white spaces at the ends of sections are also forms of typographic redundancy, allowing the channel to be less cluttered with information. More obvious is actual **repetition**, deliberate use of the same text and exhibits in more than one place—unthinkable to most publication managers. And even more interesting than the recurrence of the same text or exhibit is a practice severely discouraged by nearly every editor of technical journals: the use of art and diagrams that restate what is already in the text; saying in a graph what can be said equally well in a sentence or paragraph.

To people who worry about the short-term cost of publications, exhibits and illustrations should never be used unless they are necessary, unless they can show something that cannot be expressed in conventional sentences. Yet, I propose that well-made user publications will "back up" their sentences with pictures (and vice versa). Why? Because there are word-readers and diagram-readers, and if we want to adapt to the audience and control the transaction, we have to write for both.

Occasionally, there will be a fortunate case in which economy and usability are compatible: for example, the use of smaller text to let an entire procedure fit on one page. Generally though, most of what makes books more usable—including such items as durable, heavy paper stock and color printing—may seem expensive and wasteful at first.

In the longer view, though, the benefits in efficiency and productivity can save thousands of times what they cost. And in the broader view, money spent for more-readable documentation eliminates or contains the costs of field service, training, troubleshooting, and a variety of other expensive services. Hard-to-use documents create a demand for technical assistance, and avoiding that demand is the main cost justification for easy-to-use documentation.

Good documentation should pay for itself in enhanced productivity, improved sales, and reduced support costs. But to appreciate the economic advantage of usable documentation, there must be a leader with a sense of the diseconomies of unproductive work and a grasp of the total costs of user support, not just this quarter's printing budget.

3.6 The Ultimate Test: Reliability

The Usability Index (the degree to which manuals are free from skips, branches, loops, and detours) is not arbitrary and it is not just aesthetic. It bears directly on the cost of implementing and supporting systems. Usable documents are more reliable, *that is, less likely to fail in use.*

It is not enough that a system merely work. It must work reliably—predictably and unfailingly. (And it must be maintainable—easily serviced on those occasions when it fails to perform.) Most of the advances in system development methods, the whole repertoire called "software engineering," pay for themselves by giving us more reliable and maintainable software and technology.

Now that programmers have finally begun to think this way, it is time to get the message to the documentors, including the professional technical writers. If documentation is a system, and if each manual is a component or device in that system, then each manual should be built for reliability and maintainability as well.

But what is meant by "reliability of a manual"? In what sense does a manual fail? Can a manual be appraised by tabulating its mean time between failures (MTBF being the most popular reliability metric in engineering)? Does a manual really break down?

A manual may be said to have failed if the user/operator is unable to work because of it—if a mistake, malfunction, or interruption can be blamed directly on the manual. Failures can result, then, from omitted information, incorrect information, or ambiguous or contradictory information. Or failures can result from an inaccessible arrangement of materials that raises the effort needed to find information, leading to false starts, frustrated efforts, or improvised solutions to problems that cannot be handled with the manual.

The first main justification, then, for the Usability Index is that there is a direct connection between the number and complexity of the skips and loops in a book and the number of errors and breakdowns likely to occur. The more paths there are through a publication, the higher the chances of taking a wrong path. The more discontinuous movement through a document is, the higher the chances for a wrong move. The more choices the reader has to make, the higher the odds for a wrong choice.

This prediction applies most directly to readers with limited experience, and especially to those with modest reading skills. But no one should mistake this principle as applying *only* to people who have trouble with complicated books. Even though some users are accustomed to tangled, unreliable books, no one likes them.

Reliability can be treated as a target. Documentors who are trying to reach Reader X (the person who has trouble learning from books) had better set the target high. Those writing for Reader Y (the person who is used to complicated books and is not afraid of them) can set it somewhat lower.

For example, should the manual for an application duplicate and incorporate material from the operating system manual, or should the reader be directed to the other publication? Should certain routines that occur repeatedly be presented in full each time they occur, or should they appear once in the text, with appropriate page references elsewhere?

Documents, then, suffer some of the same problems as programs: complicated, tangled logic leads to breakdowns and then slows the process of repair. Fortunately, the techniques devised to solve these problems in computer programs can be applied with little modification to the writing of usable user documents.

Exhibit 3.6: Measures of Reliability

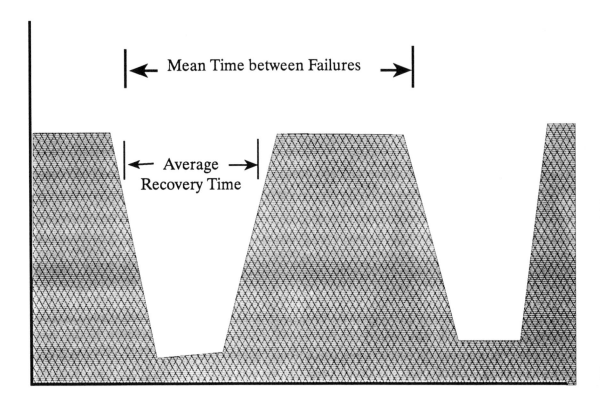

PART 2

A Structured Approach to User Documentation

4. "CULTURES": HOW DOCUMENTATION GETS WRITTEN

4.1 Two Ways to "Write" a Document

There are two broadly different ways to write a document. The first is to compose *it, crafting the sentences and paragraphs while they are being written, as would a writer working on a script. The second is to* engineer *it, preparing a series of increasingly finer specifications until, at last, a document "drops out."*

When people think of writers, the image that usually comes to mind is a stereotype of a person slaving over the sentences—few notes, no plans, no models or mockups. There are just the blank pages (or screens) and the writer's mind.

There is a similar stereotype for computer programmers: people who think with their hands on the keyboard—trial and error, inspired guesses, flashes of genius. In movies and on TV, programmers never consult a dataflow diagram!

There are, in fact, such stereotyped programmers and writers. They are usually either amateurs or professionals working on very small projects. It is when they bring this approach to expensive and complicated projects that the trouble starts.

Programming and documenting, you see, are two of the very few complicated projects that can actually be carried off in this loose, unplanned, artistic style. (The term *artistic* is not meant to imply that all or most artists work this way; rather, that is the popular conception of how they work.) No one would manufacture a car that way or build a bridge by trial and error. Indeed, computer programmers and technical writers are among the very few people I know who would, without hesitation, invest six person-months of effort on a nearly unspecified project and hope for it to turn out well.

Putting the issue somewhat differently, there are various attitudes and "cultures" that can influence the writing of user documents. As Exhibit 4.1 shows, the artist puts relatively little effort into planning. The main push is in the drafting stage—which is often interrupted for lack of ideas and inspiration. Thereafter, the biggest effort is applied to patching up the problems in the manual, a task that trails off into infinity.

Exhibit 4.1 also shows the distribution of effort for the engineer. (The term *engineer* is not meant to suggest that all engineers work this way.) Here, most of the effort is in the planning—definition, design, modeling. The draft is merely the implementation of the design, not the creation of the product. And because engineers seek out problems early and solve them while it is still cheap and easy to do so, relatively little patching is needed.

A complementary distinction between the two cultures concerns *when other people get involved.* The artists do not want to show the work until it is "ready." Usually, no one but the artist gets to review, test, or criticize the work of art until it is virtually finished. In contrast, the engineered product is discussed extensively—and criticized and revised extensively—at several intermediate stages, before the author's ego is too deeply invested in the work.

Again, the terms artist and engineer are not supposed to suggest that all artists work without planning or that all engineers are so perfectly disciplined. In truth, many professional writers prepare elaborate plans before they commit themselves to a draft and many engineers solve problems with casual trial and error—what programmers are likely to call prototyping. Rather, the purpose of the distinction is to emphasize that certain professions can be practiced with either "culture" and, moreover, that when the projects get complicated and the stakes get high, the artist should yield to the engineer.

Exhibit 4.1: Two Cultures of Technical Writing

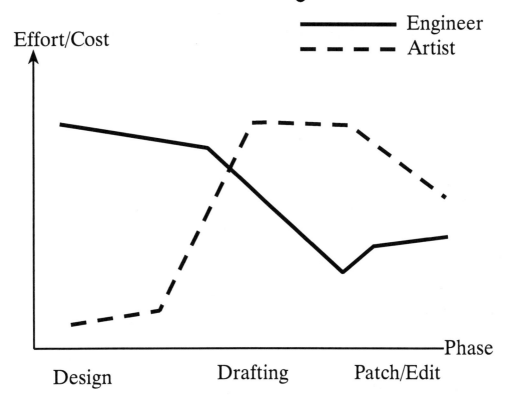

4.2 What Documentors Can Learn from the History of Programming

Programming has evolved from an informal craft into a formal branch of engineering, its emphasis shifting from coding to design. Fortunately, many of the tools developed to improve programming can be adapted to writing user manuals.

In a 1979 issue of *ComputerWorld* Robert Perron wrote that, "a comparison of programming in its early days to technical documentation in its present state yields some striking similarities." That is, people writing manuals in the 1980s and 1990s often resemble the people who wrote programs in the 1960s. They are prone to the same errors, they have the same foibles, and their products (publications) have flaws similar to those in the programs written by the earlier group.

In the late '50s and early '60s, computer programming was an exotic, if not eccentric, profession. The people working in it were drawn by aptitude and passion. They were *not* trained by the schools and colleges and they were *not* treated like ordinary white-collar employees. Programmers worked alone, like artisans, often without much supervision and sometimes without budgets or deadlines to worry about.

Two factors, more than anything else, changed the nature of the programmer's job. First, the typical program became too large for one person working alone, ending the solitary luxury of the programmer. Second, the major expense of programming shifted from inventing programs to *maintaining* them, and with that shift came the realization that most programs were disorderly, tangled, unmaintainable messes. Both these important developments led to the invention of software engineering methods and "structured" techniques, and to a redefinition of the programmer's occupation.

In today's organization, however, it is more likely to be the writer who is treated with defer-ence, who works with little supervision and not much budgetary constraint. Today, for example, *most companies have no idea what it costs to write a page of user documentation.*

But just as complexity, size, and maintenance problems made the old way of programming obsolete, so are they making the old way of writing manuals obsolete. Today, writing must be managed, budgeted, scheduled, and done *by teams of writers working in parallel.* And if documents have to keep pace with systems that are revised every few months, the manuals have to be modifiable. For the most part, then, the era of the artist-documentor is over.

And what lessons have the programmers learned that the documentors should also learn? First, the single most important principle of software engineering: *the cost of detecting and correcting a problem rises exponentially as a function of how late in the development cycle the problem occurs.* That is, what costs a few minutes or a few dollars to fix at an initial planning session can cost hundreds during design, thousands during implementation, and tens of thousands during distribution and operation.

Programmers have also learned the psychological implication of this principle: *The more costly and complicated a needed change, the less likely it is to be made*, or made properly. So the essence of structured methods is to develop products in a such a way that problems and flaws appear as early as possible.

To become an engineer, then, either a programmer or documentor must adopt an attitude that may come hard at first: *an eagerness to find errors.* Usable and reliable technology is the result of testing, and the function of testing is to make things *fail.* Anyone who hopes that the test will show no flaws, that the specification will generate no arguments, that the outline will raise no questions—that is, anyone who hopes that errors will come up later (rather than sooner)—is asking for expensive problems and poor quality.

In sum, what documentors must learn from the history of programming is the craft of *top-down design and testing:*

1. The sooner an error or problem is detected, the cheaper and easier to correct it. Therefore, privacy and informality in the early stages of a manual are quite expensive.

2. The most serious problems in a complicated product are usually in the connections and interfaces, not in the units or modules. There-fore, the cost-effective way to develop a manual is to build it top-down, to assure the right mix of documents and the right content and sequence within each document *before the draft.*

3. Unless a project has been designed top-down, it may take longer for several people to do the job than for one person working alone. Therefore, when documents must be prepared on an accelerated schedule, they must be written to a detailed, top-down model.

4. It usually costs much more to maintain and support a complicated product or system than to design it simply in the first place. There-fore, the claim that there is not enough time and money to develop high-quality, maintain-able manuals is nearly always false.

4.3 Goals for an Effective Process

What is needed is a documentation process that raises the level of debate *and improves the suitability and appropriateness of the available documents; reduces the skips, jumps, and detours; enhances clarity, readability, and reliability; and makes the publications easier to maintain.*

Even those firms that have begun to conduct formal, rigorous usability tests of their documents will soon learn that errors in a complete draft are far more recalcitrant than errors in an outline. Consider the analogy with those firms that do aggressive unit tests of their program modules but just cannot seem to integrate the tested modules later on. What these firms seem to overlook is that program modules, like book modules, must be integrated *before* they are written, not after. And that the most agonizing problems in writing or reading documents are in the links and connections, the interfaces, not the individual units or pages.

A documentation process that learns the lessons of software engineering will achieve the five goals listed below.

It will improve the "fit" between user documents and the needs and convenience of the users. The method must include a way of aligning the material to be written—the information products—with the users of those products. In other words, the process must be driven by the particular characteristics of the users and operators and their peculiar interests in the system, rather than follow a one-size-fits-all standard for user publications. More simply, the process must recommend ways to define a logical mix of information products and services, a *user support envelope.* Furthermore, the proposed contents of this mix must be *testable as a proposal.* That is, it must be possible to review the plan and find strategic errors in the making, well before the plan is turned into manuals and disks.

It must reduce the skips, jumps, and detours. Even though modern word technology makes it easier to move blocks of text around, it is still inescapable that once a document is completely drafted, it develops an inertial resistance to structural change. To be effective, any technique used for documentation will, necessarily, expose structural errors *before* the inertia of the draft takes hold. An effective documentation process will generate a series of increasingly more detailed models of the product. And these models—which appear between the outline and the draft, the interval when the artistic writer usually works alone—will be testable against clear measures of usability.

It will allow writers to work in teams and in parallel. The most common excuse for inadequate documentation is the claim that preparing it would delay the delivery or implementation of a system by several weeks or months. But this excuse is, rather, a clear indication of the need for techniques that will allow user documentation to be written by teams of people, working on well-defined chunks of the publications, *in parallel.*

Note that without the right method, having writers work in teams can actually *slow* the process; with the wrong approach, two writers will take two or three times longer to write a book than one writer! An effective documentation process will organize the work into a set of manageable parcels, capable of *independent execution*, with costs and schedules that can be predicted and controlled.

More specifically, it will "decompose" the large job of writing into a set of small jobs, tiny documents, of easily estimated size and cost. Indeed, in some methods, each tiny document is of about the *same* size and cost. Furthermore, all the links and interfaces between the tiny documents have been defined, explicated, and tested in the model. Thus, each of the small pieces can be written independently, without consulting the authors of the other pieces—so long as each author has access to the model of the whole publication.

It will enhance the clarity, readability, and reliability. The flaws in the draft are, of course, important, but the goal is to solve every strategic problem and correct every structural flaw *before* the draft is composed. In that way, what will remain to be corrected in the draft are precisely those problems that lend themselves to editorial improvement: incorrect claims about the system, minor technical changes, unclear sentences, ambiguous paragraphs, cluttered or confusing illustrations.

Furthermore, an effective documentation process will include *formal* standards for editing and will not rely on the artistic, intuitive, "stylistic" preferences of one person. For example, the prompt "Press F(4) to continue" is a backwards, unreliable sentence. A sound documentation process will flag and correct this sentence, whether or not the editor finds it personally objectionable, and whether or not any reader has trouble with it in a test.

It will generate publications and products that are maintainable and modifiable. Well-made documents will not have to be revised and supplemented as often as ill-made documents. But, when they *do* need revision, the process will be more rational and manageable.

5. DEVELOPING DOCUMENTS: A STRUCTURED APPROACH

5.1 What *Structured* Means

The term structured *can be applied to user documentation in two main ways. First, the process for developing user manuals is characterized as a "structured process." Second, the publications themselves are often called "structured documents." Unfortunately, the word* structured *is used so often and so casually these days that it is necessary to pause and define what it means.*

When I use the word *structured*, I am not referring to its overworked conversational meaning, in which it is a loose synonym for disciplined or organized. Rather, its sense is the one it has when used by computer scientists or software engineers in such expressions as *structured analysis, structured design,* and *structured programming.*

In all three uses, **structured** refers to a certain process or method, well put in the following definition of structured analysis:

A formal, top-down decomposition of a problem or process into a model that offers a complete, precise description of what the problem is . . .

— Sippl and Sippl, *Computer Dictionary & Handbook* (Indianapolis: Sams & Co.), 1980, p. 529

First, structured analysis is **formal**, that is, explicit, and rule-abiding. A process cannot be considered structured if it is intuitive, private, or conducted without rules or guides. (In practice, formal methods compel us to generate **evidence**, records that prove we have honored the rules.)

Next, it is **top-down**, which means that it starts with the biggest picture possible, the whole system, with all its interfaces, and adds overlays of detail in its successive stages. And at each consecutive level it is tested, using "stubs" or dummies for the processes below that level.

Many people confuse top-down with the next key term, **decomposition** (disaggregating big things into smaller things). Although structured analysis requires decomposition, it first requires a representation of the entire system. In structured technology, we know that the parts fit into the whole before we define the insides of the parts.

The next key word is **model**. Put simply, in structured methods we build models of a product before we build the product itself. Why? Just because it is much cheaper to build and change models than to change the finished product.

After structured analysis comes structured design:

The art of designing the components of a system and the interrelationship between those components in the best possible way. Or, the process of deciding which components interconnected in which way will solve some well-defined problem.

— Yourdon and Constantine, *Structured Design* (Englewood Cliffs, NJ: Prentice-Hall), 1979, p. 8

Notice that a product designed this way has only two things in it: components and the relationships between them—modules and interfaces, nodes and edges, units and links. And because there are only these two kinds of entities, it is usually possible to describe a structured product or system with only a simple diagram containing blocks or circles for the modules (nodes, components, units) and arrows or lines for the connections (linkages, interfaces, edges).

The reason for making such diagrams—especially in the planning of a complicated document—is to find flaws and problems while it is still cheap and easy to correct them. And the ultimate benefit of such a design is later, in the maintenance phase, where all changes will consist simply in replacing or adding one small module or unit, and where the effects of making that change will be predictable from a study of the design.

The same structured methods used to make programs and systems more cost effective and *maintainable can be applied directly to the job of designing and writing user documentation,* and with similar benefits. Further, if the process is structured, then the products—the publications—will also be structured. They will consist of many small components (modules) connected in a way that makes the book as usable and maintainable as possible.

Exhibit 5.1: Maxim from *Structured Design*

It is always easier (and cheaper) to create two small pieces to do the same job as the single piece.

...Yourdon & Constantine

5.2 What *Modular* Means

The most conventional definition of a module *calls it a small, independent functional entity, a component of some larger entity. Well-made modules are cohesive and predictable; well-designed modular products are free from excessively complicated couplings across modules.*

Modularity is elusive: Designers and engineers can "feel" modularity when they get close to it, but are hard put to define it operationally. Consider the parts of the definition in reverse order.

Modules are functional. Modules are not just parts of something larger; they are *functional* parts. A module performs some task, it converts data from one form to another, more usable form. Moreover, well-made modules usually perform a whole task, for example, sorting all the accounts payable in a file according to their age. A well-made module is also predictable: The same inputs arriving under the same conditions will generate the same output; there is no "internal memory" in the module that would change the input/output patterns.

Modules are independent. Because modules are not dependent on their context, a module with a particular function will perform that function in more than one setting. Any module can become part of a library of reusable modules; eventually, designers can create systems or products from the catalog of available modules.

Modules are small. The least precise part of the definition refers to their size. To say that modules perform only one function fails to limit them precisely. Long arguments about whether something is one module or more than one are usually unproductive. Most people who work with structured methods limit the maximum size of a module. In data processing, the limit is usually a certain number of code statements; in publications, it is a certain number of pages.

Indeed, one of the interesting parts of devel-

oping modular products is playing with the size of the modules. As modules get larger, they get less cohesive (have more than one function); as they get smaller, though, the couplings and connections become more complicated. In modular publications, these couplings manifest themselves as references to other pages in the book. And a central argument of this book is that these sorts of design decisions—such as trading-off module cohesiveness for inter-module complexity—can be applied directly to the development of more usable user documentation.

The first mature attempt to treat documentation this way was the invention of a group of publication engineers working for the Hughes Aircraft Corporation. Their process, a form of "storyboarding" adapted from the motion picture industry, and their modules, two-page spreads, are described in their seminal work on the subject:

Tracey, J.R., Rugh, D.E., and Starkey, W.S. *STOP: Sequential Thematic Organization of Publications*. Hughes Aircraft Corporation: Ground Systems Group, Fullerton, CA, January 1965

A more recent summary of Tracey's reflections has also been published:

Tracey, J.R. "The Theory and Lessons of STOP Discourse," *IEEE Transactions on Professional Communication*. PC-26 (2) June 1983: 68-78

Modular manuals benefit not only the readers of manuals but also the developers and writers. Working from modular outlines, designers are able to predict the size and cost of publications at the same time they are testing them for readability and accuracy. Furthermore, by breaking the long, complicated process of writing into a set of small, independent tasks, firms can apportion the writing assignments to a great many people who can work in parallel, independent of one another.

Modular manuals are also a boon to "authors"—all those people we usually call on for raw input to the manuals. In the modular manual, these people can be transformed into "first drafters," each knowing exactly how much to write and exactly what points to cover.

Even writers working alone as "artists" benefit from modularization. They can work in short bursts, knowing that the little pieces will ultimately fit together well.

The modular approach especially benefits those who manage writing or supervise publication. Planning, writing, editing, and producing by module enhances the control of the person in charge.

And perhaps most important, effectively designed modular documents are the most readable and "friendly" technical publications imaginable. Indeed, the reactions of readers to modular publications have done more to sell the concept than all the arguments by consultants. Although there are some technical writers who dislike modular manuals, I have never yet met a reader who does.

5.3 Good Documentation Is Modular

A modular document, like a modular system, is made up of many small, functional, independent units, or modules. Given the right planning and design, a modular document is far easier to write and maintain than a traditional, "monolithic" document. Most important, though, modular documents, because they are free of many of the flaws that make publications unusable and unreliable, are much easier to read.

The book you are reading is a modular publication, designed after the style invented by J.R. Tracey and associates at the Hughes Aircraft Corporation in the early 1960s. Put simply, such a document is conceived, planned, and outlined as a series of small self-contained units, each containing all the words and exhibits needed to grasp a single concept or theme.

The most apparent innovation in this technique is the consistent use of two-page spreads as the basis of organization. That is, with rare exceptions, all the material in the book is pre-

Exhibit 5.3a: Shell for Hughes Module (STOP)

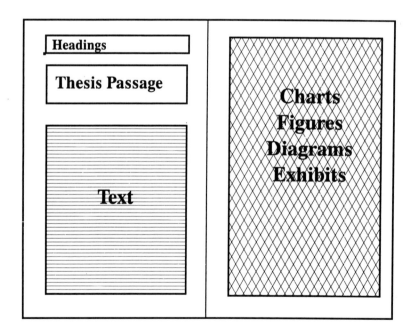

Headings

Thesis Passage

Text

Charts
Figures
Diagrams
Exhibits

sented within modules that contain two facing pages. Every figure or table that needs to be seen *can* be seen, without turning or riffling pages. And any concept too big to embrace in one module is "leveled" or "chunked" into a series or hierarchy of modules.

Modular publications are planned top-down. That is, before anyone writes a whole page of draft, some designer or team of designers has decided precisely what modules are needed, in what sequence. Moreover, there is a "spec" for each module, defining its scope and content.

Writing by module reduces the burden on writers. If a document is broken into two-page (or one-page) chunks, it can be written in short bursts of time—which is how most people must write. Moreover, the modules are reusable in other publications.

In a modular book, it is far easier to know whether the book is current. Maintaining a modular book consists in replacing inaccurate modules or adding new ones (whereas in a traditional, monolithic publication, no one is even sure where the errors are).

And, finally, modular books are easy on the reader. If modules are limited to two facing pages, or one page, or one screen (panel), the most common problem in using technical publications—searching for disjointed text and figures—is solved.

Modular publication is one of those rare practices that makes life easier for both writers *and* readers.

Exhibit 5.3b: Spec for Two-Page Module

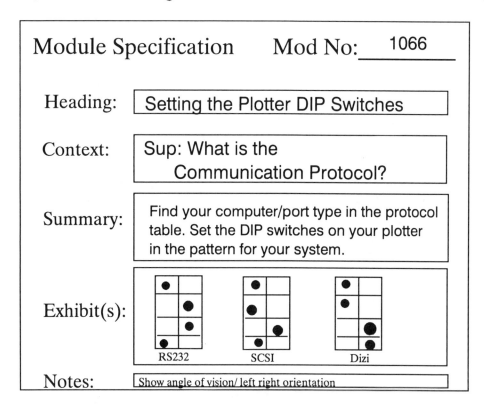

5.4 Overview I—A Dataflow Diagram for Developing Documentation

Like a system, user documentation has a life cycle: analyze *support needs,* outline *each information product,* storyboard *each product,* assemble *text and draft,* edit *for correctness and readability,* test *with representative users, and* maintain *in the face of system changes and revealed errors.*

Documentation is never really finished. As the application or product changes—as its "bugs" manifest themselves—there is a need for more explanation and teaching. Thus, the development of documentation is cyclical. There are seven main tasks:

Analyze—Convert product descriptions and F-specs into a user support plan and its associated list of information products and services (the user support envelope).

Outline—For each information product, develop a series of increasingly refined outlines (topical, substantive, modular), rich enough for review and testing.

Storyboard—For each entry in the modular outline, prepare a module specification and mount the specs in a "gallery" or storyboard, which is reviewed and adjusted by all affected people.

Assemble—Assign, collect, write, and reuse the material called for in the storyboard, using project management techniques that allow writers to work in parallel.

Edit—Correct and improve the first drafts to eliminate technical errors and also to improve their readability and clarity.

Test—Conduct formal, controlled tests with representative users and adjust the publication as needed.

Maintain—Immediately begin surveillance of the documentation to search for strategic misalignments of books and audiences, structural or organizational problems, missing explanations, lapses of style, and, of course, any technical errors.

All seven phases are logically necessary. Organizations that skip some, especially the first two or three, may produce documents with all the expensive flaws of underanalyzed computer applications and carelessly engineered machines.

Exhibit 5.4 Dataflow Diagram for User Documentation

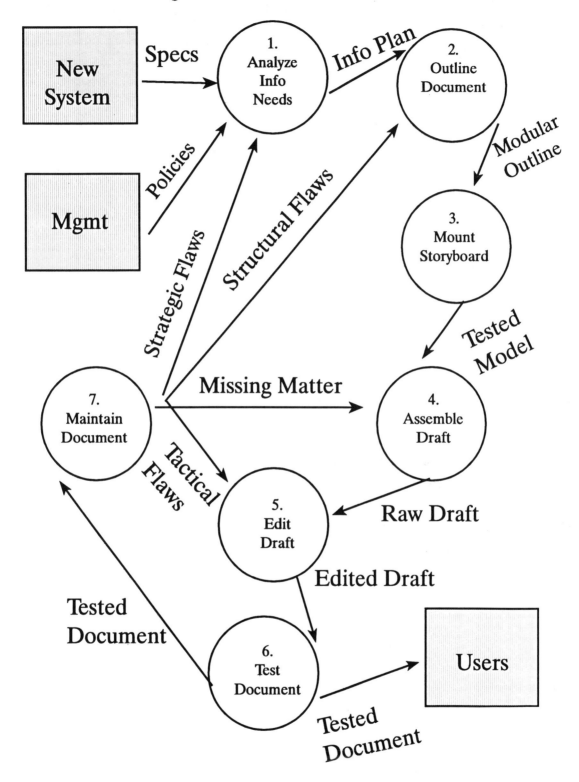

5.5 Overview II—A Work Breakdown for Developing Documentation

The table below and on the facing page shows a breakdown of the tasks needed to develop usable user documentation and related information products.

Exhibit 5.5 lists tasks that must be completed in high-quality user documentation. Although it does not include every task (such as cost estimating or printing), it *does* include every activity needed to realize the structured method advocated in this book. That is, anyone wishing to skip some of these tasks or perform them in another sequence must have a persuasive justification for doing so.

Note also that the list does not address the question of *who* should perform each task. Because there is so much much variety in the staffing and organization of firms and agencies who write user documentation, it is impossible to say who should do what. Rather, the next six chapters discuss each of the tasks in some detail and advise on the kinds of skills needed for each.

Exhibit 5.5: Work Breakdown for User Documentation

I. Analyze Support Needs

1.1 Form documentation/support team
1.2 Review the product description/specs
1.3 Prepare task-oriented topic list
1.4 Prepare hierarchical user list
1.5 Compile the *User:Task Matrix*
1.6 Iterate the matrix until satisfied
1.7 Prepare preliminary pubs plan or support envelope
1.8 Review, test, and adjust

II. Outline a Document

2.1 Choose an information product
2.2 Assign design team
2.3 Extract topics from the matrix
2.4 Prepare topical outline
2.5 Prepare substantive outline (optional)
2.6 Prepare modular outline

Exhibit 5.5: Work Breakdown for User Documentation (continued)

III. Storyboard the Document

3.1 Prepare a spec for each module
3.2 Schedule, set-up storyboard session
3.3 Mount the gallery of specs
3.4 Conduct reviews with likely contributors ("authors")
3.5 Conduct reviews with representative users
3.6 Incorporate necessary changes
3.7 Secure official approval and "freeze"

IV. Assemble the Draft

4.1 Assign "authors"
4.2 Retrieve archive materials
4.3 Coordinate, collect drafts

V. Edit the Text

5.1 Edit drafts for conformity with specs
5.2 Edit drafts for clarity and readability
5.3 Edit drafts for consistency and conventions
5.4 Edit drafts for technical accuracy/currency

VI. Test for Usability

6.1 Develop test protocols/data sets
6.2 Schedule test sessions
6.3 Select, brief subjects
6.4 Conduct entry interviews
6.5 Run test (observe unobtrusively)
6.6 Conduct exit interviews
6.7 Interpret results
6.8 Revise publications and retest until standard is met

VII. Maintain the Document

7.1 Assign responsible maintenance unit
7.2 Create maintenance files
7.3 Receive responses from users
7.4 Seek out responses from users
7.5 Conduct further tests
7.6 Modify publications

- Strategic realignments
- Structural reorganizations
- Additional materials
- Editorial improvements

6. ANALYSIS: DEFINING WHAT PUBLICATIONS ARE NEEDED

6.1 Preparing for Analysis

Analysis begins with the appointing of the members of the documentation team: first, an expert on the technology, second, an expert on the uses of the technology, and, finally, a coordinator to bring these two perspectives together.

Most firms get nowhere with their documentation problems until they empanel a team whose mission is to name and describe the documents to be developed.

Deciding what documentation is needed is too important a decision to be made casually or by default. It is also too important to be made by one person. Rather, defining what documents are needed calls for at least three perspectives:

- the technology expert
- the application (or user) expert
- the documentation coordinator (user support technologist)

The technology expert is the team member who knows the most about the design and inner workings of the system. Known variously as a systems analyst, lead designer/developer, project manager, or just engineer, the technology expert must speak for the system. If the system has already been developed, he or she must be most knowledgeable about its features and characteristics. If it is about to be developed, he or she must be in charge of the functional specification or general design.

The application expert (often referred to simply as the user) must know what the system is for. Candidates for this position on the documentation team include users, end users, operators, manufacturers, technicians, auditors, marketing managers, trainers, and customer relations people. The task of the application expert is to remind the team members—as often as necessary—that the system will have to be used and operated. And that the users and operators (and supervisors, administrators, and even salespeople) usually want to know *how to work the system, not how the system works.*

Often, these two perspectives will conflict—just as users' requests and analysts' responses often conflict. Thus, the third member of the team, the coordinator, must manage the conflict and forge a consensus. The coordinator (known variously as documentor, business systems analyst, liaison, quality assurance rep, technical writer, support specialist, or even publications engineer) must produce the actual plan. He or she must listen to the others, follow some procedures that will be discussed later, and produce the information support plan.

Note that the person responsible for documentation planning is a manager/coordinator involved early in the life cycle, rather than a copy editor brought in to clean up untidy drafts. Note also that the job of defining documentation needs cannot be left either to the technical expert or to the application expert alone; in general, neither sufficiently appreciates the other's point of view. And in practice, they often find it hard to communicate.

Ideally, there should be three members on the team, one from each category. There may be more if the system is complicated or has an unusual mix of users. Be careful, however, that the appropriate power prevails if there is more than one technical expert.

Also beware of the two-person documentation team, which, for the sake of "efficiency," suppresses legitimate conflict. And be especially cautious if the support plan is the work of only one person.

Exhibit 6.1 Members of Support Planning Team

Perspective	Candidates	
■ Technology	■ Analyst/Programmer ■ Lead Designer ■ Project Manager ■ Hardware Engineer ■ Software Engineer	
■ Application	■ User ■ Operator ■ Trainer ■ Technician	■ Auditor ■ Marketer ■ Supervisor ■ Consultant
■ Coordination	■ Editor/Writer/Documentor ■ Business/Functional Analyst ■ Liaison/Coordinator ■ Quality Assurance Rep ■ Publications Manager	

6.2 User Manuals in a Support Context

There are certain fundamental questions about the content of a publication that cannot be answered without defining the larger set of documents and information products of which it is a part.

The proper way to begin user documentation is to define a set of information products (books, reference cards, videotapes) and then to define a specific function and scope for each item in the set. Why? Because to define what a thing is, you must also define what it isn't. The surest way to clarify the purpose of a publication is to contrast it with other adjacent publications.

The systems approach to a problem consists in viewing it as part of a larger problem. Before we can know what to put into a particular book, we must know why there are any books at all, what they do as a group, and what they do as individuals.

Indeed, the most appropriate way to define a set of user documents is to think of them as part of a larger set of items called information products, including not only publications but also audiovisual products, online tutorials, and the whole range of teaching and reference media.

Furthermore, the most appropriate way to define the needed set of information products is to view it as part of a still larger entity called user support, which contains not only information products but a full range of user services. (See Exhibit 6.2.)

Notice also that there are even trade-offs between the quality of information products needed and the quantity of services needed; high-quality information products can reduce the need for training, consulting, and maintenance. In fact, that is a main cost justification for investing in user documentation.

Put simply, the time to decide the scope of a particular publication is *not* during the writing of the outline and certainly not during the writing of the draft. The time for definition is *before the outlines are written.* The time to argue about whose information needs will be served is at the beginning; the time to argue about whether two publications will overlap is before either of them has been outlined; and so forth.

Yet, as obvious as this principle may seem, most writers of manuals ignore the issue. Like programmers eager to produce some code, the documentors are eager to produce some text.

And the consequences are the same. The finished draft, like the coded program, develops inertia; its author becomes its defender. What users or customers need has far less influence than what has already been written and paid for.

Right now, there are hundreds of skillful writers struggling with undefined and misconceived publications. Unfortunately, these writers think that their problems are *within* the publication. Actually, the problem is strategic: the lack of an information support plan.

It avails us little to be competent writers if we write the wrong manual. And the only way to be sure of what a manual is, the only way to know what to include and what to "include out," is to differentiate each product from the others in the set.

Exhibit 6.2 Manuals in a Support Context

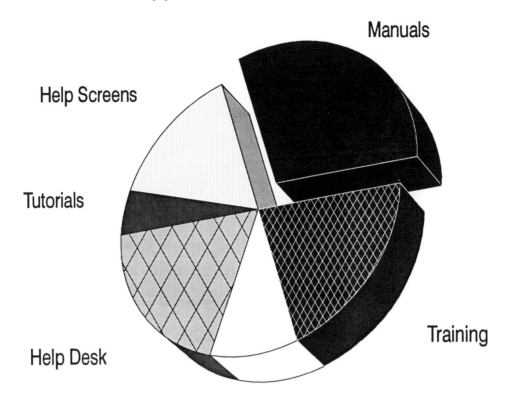

6.3 Using the "Universal Task Architecture"

The consensus today among technical communicators is that good documentation is task-oriented. That is, it helps readers perform specific tasks; it does not describe the product in meticulous detail. A useful framework for generating lists of tasks is IBM's "universal task architecture."

The aim of customer documentation has changed dramatically in the past decade. In 1980, good user documentation was called comprehensive; in the 1990s it will often be called minimalist. In an era when systems technology was unfriendly and opaque to all but the most sophisticated users, the aim of the user documentation was to describe everything about the system.

Today, the emphasis in computer and communications technology is on ease of learning and use. Documentation is to help people do their jobs. Most users *do not want more information than they need*. Indeed, one of the most insistent documentation problems is the inability of competent readers to *find* what they want in an unnecessarily detailed publication.

Task analysis, then, has two purposes: first, to reshape our notion of documentation away from product description and toward task support; and, second, to differentiate the support needed by different audiences performing different tasks.

Although there are many possible frameworks for developing a list of tasks, IBM's "universal task architecture" (IBM Publication ZC28-2525) is a nonproprietary scheme enjoying wide popularity among writers of software documentation. Briefly, the universe of tasks contains nine broad (and overlapping) categories:

Evaluation—Considering information that will influence the selection, acquisition, and purchase of systems; consulting documentation that advertises benefits and guides choices

Planning—Carrying out activities that precede the arrival of the product/system

Installation—Setting up and configuring the product

Resource Definition—Making adjustments in the environment or associated technology, needed to accommodate the new product/system

Operation—Starting and stopping and performing the basic manipulations and transactions of the product; input and interface conventions

Customization—Setting the defaults, or altering those that are shipped with the product

Application Programming—Building chains of transactions and operational elements into programs and processes that do useful work; creating macros; programming

Program Service—Assessing and solving technical problems (for the customer); field maintenance

End Use—Performing specific occupational tasks and activities, peculiar to the customers' profession or assignment

For most computer and communication products, there will be a list of 100 to 200 supported tasks, some of them requiring many pages, others only a sentence or two. Note that the easier a system is to install and operate, the less information support is needed. Indeed, when designed for ease of use, some systems eliminate many of the error-prone tasks that users have to perform and, thereby, reduce much of the documentation needed. Consider, for example, the very different documentation burdens for these two tasks:

- Writing Patches to Alter Printer Characteristics
- Starting the Printer Setup Program

Exhibit 6.3: The "Universal Task Architecture" (IBM)

■ **Evaluation**

■ **Planning**

■ **Installation**

■ **Resource Definition**

■ **Operation**

■ **Customization**

■ **Application Programming**

■ **Program Service**

■ **End Use**

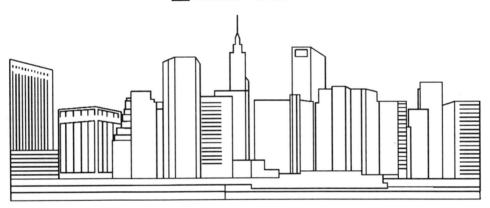

6.4 Listing Features and Topics

The technology expert on the team analyzes the components and features—the topics—that need to be written about. Although there are countless schemes for categorizing the aspects of a system, and although a task-oriented system is best, the particular approach is less important than the completeness *and* fineness *of the analysis.*

In analyzing the documentation needs for a particular product, a critical task is to decide just what the product is. Just what is worth knowing about it.

The topic analysis (or sometimes functional analysis) is the job of the system expert on the documentation team. Although the breakdown will inevitably be influenced by the other members of the team, it is still the system expert's job to describe the structure and morphology of the system itself.

Systems can be described variously by talking about their physical components, their design, their technology, their operations, their applications, or their benefits.

But since the 1980s, the talk in documentation circles is of task-oriented manuals, which are organized according to the tasks performed by the intended reader. Task orientation—in contrast to product orientation—is an application of what the instructional technologist calls "skills analysis" and what some social scientists call "activity accounting."

As will become clear in a while, the best way to eliminate loops and detours from a manual—to raise its usability—is to make it task-oriented for a well-defined audience. (Remember, though, that even if this initial breakdown of topics is not task-oriented, there are still opportunities later in the documentation process to incorporate task thinking into the design).

There are several orthogonal paths of attack in defining features and topics. Although task breakdowns are nearly always more useful for documentation than physical breakdowns, any scheme will do as long as the analysis is fine enough.

How fine? The topics must be small enough and clear enough so that team members can ask the following general question: Is Topic T necessary or important for Reader R? Yes or No? If the topics are defined too broadly or vaguely, then the analysis must be refined.

Exhibit 6.4 Tasks Generate the Topic List

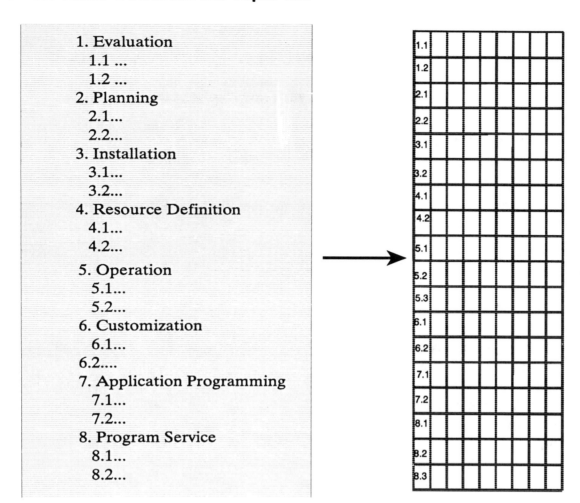

1. Evaluation
   1.1 ...
   1.2 ...
2. Planning
   2.1...
   2.2...
3. Installation
   3.1...
   3.2...
4. Resource Definition
   4.1...
   4.2...
5. Operation
   5.1...
   5.2...
6. Customization
   6.1...
6.2....
7. Application Programming
   7.1...
   7.2...
8. Program Service
   8.1...
   8.2...

6.5 The Concept of a User-Audience

In this process, an audience is a cluster of people with a common information deficit. An information deficit is defined as "what one needs to know, reduced by what one already knows." Thus, people are differentiated not only according to the unique requirements of their tasks or occupations (need), but also by their background and experience (current knowledge). Additionally, some audiences are further divided by their attitudes toward publications.

The analysis and listing of audience has three steps:

Division by Occupation—Users/readers are grouped according to their professions (e.g., accountants), job titles (e.g., director of operations), or particular mission (e.g., field installers). Sometimes this level of analysis is sufficient.

Division by Experience—Within occupational categories, there may be important differences in the type or quantity of experience, background, training, and so forth. These differences often affect the documentation requirement by altering what the users already know. At this stage we want to assess, for example, whether the accountant has experience with PCs, or whether the PBX operator has used PBXs before, or whether the technical writer is already familiar with desktop publishing terminology.

Division by Book Skill—Many people who are presumed to be poor readers are, in fact, adequate readers but poor users of books (Reader X). That is, in any task demanding the use of a complicated publication, they are likely to have trouble. Thus, in planning some document sets, it is important to further break apart occupation/experience categories into Reader X and Reader Y (persons skillful with books). The former will not be well supported with books, which may affect the support plan.

The list of audiences consists of the "leaves" on the audience tree. That is, if clerks have been divided into experienced and inexperienced, they become two audiences. And this expanded list is transcribed to the horizontal axis of a matrix.

For most systems and products, the list contains between 5 and 10 audiences. For widely used consumer products, however, like telephones, there might be 15 or 20 user audiences with important differences in their information needs.

It is a serious strategic error to write documentation as though it were one compendium of material aimed at a universal audience. For now, we want the "worst case": the finest possible breakdown of users and readers.

Exhibit 6.5: Audience Analysis

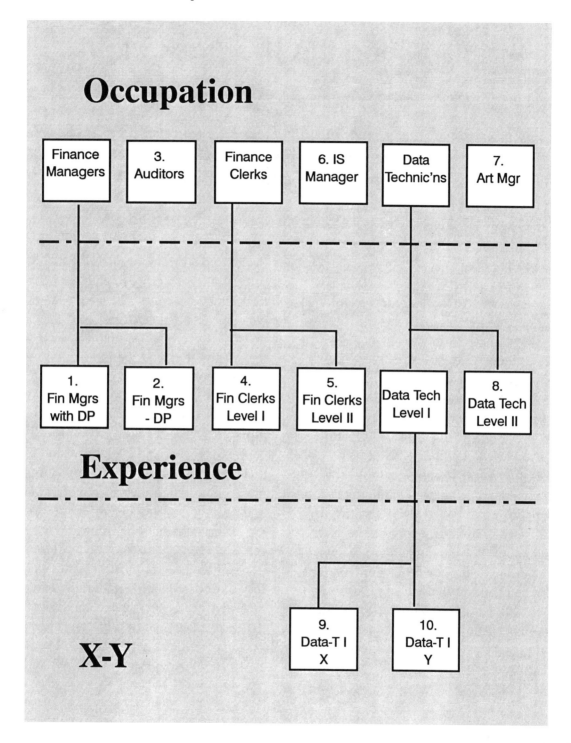

6.6 Forming the User:Task Matrix

With the two breakdowns completed, the document coordinator prepares the User:Task Matrix. The team as a whole then analyzes the intersections of topic and audience, indicating the points at which particular topics affect particular audiences.

The team transfers the two breakdowns to the User:Task Matrix and decides which topics are of interest to which users.

The decision is usually an easy consensus. When there is disagreement, the safest solution is to include the disputed topic. (In general, the best way to resolve choices about documentation needs is to provide more, rather than less, information.)

In Exhibit 6.6, the topics are refined enough to allow simple yes/no choices. That is, each topic is small enough so that a particular audience needs to know all of it or none of it.

Interestingly, a matrix very much like this one is often prepared in corporate training departments. Sometimes it is called a skills matrix or task matrix, used to define the training needs of various audiences. Indeed, if manuals are developed from this analysis, they often work as training documents as well. Instead of the usual practice—in which trainers extract thoughtfully chosen segments from cumbersome manuals— the manuals are themselves well designed for training. How often have users preferred their training materials to the company's "real documentation"!

The matrix—the process of building it and arguing about its content—is *not* just another of those time-consuming planning tasks that cause programmers to grow impatient and long to get back to their coding. The emerging pattern of checkmarks suggests the shape of several documentation products. Without it, there is a high probability that documentors will write the wrong books.

Furthermore, the matrix often shows that people with presumably different interests have remarkably similar information needs. (Data center managers and data entry clerks often receive similar checkmarks, for example.) Or, more important, it demonstrates that certain audiences have been neglected, or swamped with irrelevant information, or lumped together with readers whose needs are quite different.

Think of the long list of topics as the inventory of documentation materials; think of the list of reader groups as the consumers of those materials. The goal of this analysis, then, is to decide how the materials should be "partitioned" for the convenience and needs of the consumers.

Exhibit 6.6 Excerpt from NCR Matrix

---OPERATION---	Sales personnel	Customer services personnel	System integrators	Application programmers	Customer management	Product advocate	System administrator	PBX operator	End user (subscriber)	End user (nonsubscriber)
47. Basic operation of the Voicenter hardware. (switch settings, flex disk, streaming tape, peripherals)			●	●			●			
48. How to start and stop the system and how to start and stop voice-processing applications.			●	●			●			
49. How to add/delete/change subscriber mail boxes.			●	●			●			
50. How to add/delete/change public mail boxes.			●	●			●			
51. When to generate each of the statistical reports on system usage.			●	●		●	●			
52. How to generate the statistical reports.			●	●			●			
53. How to interpret the statistical reports.	●		●	●		●	●			
54. How to initialize streaming tapes and format flexible disks.			●	●			●			
55. When to back up the hard disk to streaming tape or flexible disk.			●	●		●	●			
56. How to backup/restore the hard disk to/from streaming tape or flexible disk.			●	●			●			
57. How to route messages from the public mail box to a subscriber's mail box.							●	●		
58. How to handle non-subscribers who request operator assistance.							●	●		
59. How to recognize and recover from a general error condition.		●	●	●			●			
60. How to recognize and recover from an NCR Voice Mail error condition.		●	●	●			●			
61. How to generate a directory of mail boxes.			●	●			●			
62. How to archive a message on disk to streaming tape.			●	●			●			
63. How to create/modify distribution lists.			●	●			●			

NCR Voice Mail end customers *

* Value-added resellers and sophisticated end customers

6.7 Interacting through the Matrix

Through the matrix, the team assesses who needs to know what. *If the topics and audiences are defined precisely enough, the team can make binary choices: Does User U need support on Topic T, Yes or No? Usually, the first analyses are not precise enough, so there will probably be two or three passes.*

The purpose of filling in the User:Task Matrix is to raise the level of debate about information needs, thereby reducing the number of users and customers who find their publications unsuitable or inaccessible. In other words, it is the not the matrix itself that is essential but the discussions needed to fill it in.

As Exhibit 6.7 shows, the process of filling in the matrix is diagnostic and iterative. The team stays at it until the map of YESs corresponds to the true information territory.

Is there a plethora of YESs? Is the matrix nearly full of them? Possibly every audience's information needs are the same; more likely the topics are too broad to differentiate. (Occasionally, the audiences are too broadly defined, but this is rarer.) Typically, the first breakdown of

Exhibit 6.7: Iterations in Developing the Matrix

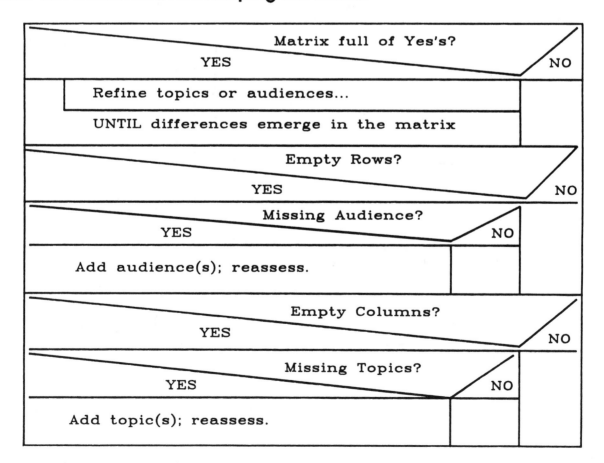

topics will need one or two revisions until patterns of difference begin to emerge.

Caution: Resist the temptation to accept a matrix that is nearly full of YESs. The central aim of this task is to break the tradition in which one compendious manual is presumed to serve everyone (The User Manual). Although it is certainly possible that, for some applications, the audience will be composed of like-minded people with identical backgrounds and interests, that should not be the default. Rather, the goal is to underscore differences in information needs and disagreements about those needs.

Are there empty rows? Probably there is an audience, or an audience subgroup, omitted from the matrix. The team should search for the missing audiences. If they find none, they may conclude that certain tasks/topics are of interest to no one. (Do we need to write about them, then?)

Are there empty columns? Probably there are omitted topics or subtopics. The team should search for the additional topics. If they find none, they may conclude that certain user audiences have no information needs. (Do we need to document for them anyway?)

Again, there is no magic in the matrix itself. Rather, the benefit is in the search for topics small enough that they enable adaptation to the audiences. And the seriousness of the debate determines the usefulness of the resulting analysis.

6.8 Assembling the Plan

The team studies the matrix of tasks and users for patterns or clusters that mark off individual documents. At one extreme, the planners may decide to prepare a single, encyclopedic reference manual—with the matrix as a guide for the reader. Or they may decide to have separate volumes for each audience or each functional cluster of tasks. Somewhere between these extremes lies the best solution, the result of trade-offs.

As in defining the boundaries of the systems, there are many decisions and trade-offs—often arbitrary rulings—involved in defining and delimiting the documentation products.

Various factors are played off against one another as the members of the team devise the most cost-effective mix of books, online facilities, audiovisual materials, and other information products. Of course, few organizations ever produce more products than they planned, tending to partition toward the encyclopedic end of the continuum.

As Exhibit 6.8 shows, there is something to be said for and against each strategy. Individualized documentation, up to the point of having separate versions of the manuals for 10 or more audiences, has many communication advantages. It results in publications that are tailored precisely to the interests of the readers, thereby freeing them from searches and detours to other publications. It generates shorter, more specialized publications, which can even have prestige attached to their ownership—something not possible when everyone has the same version.

Short, individualized manuals also protect the security and confidentiality of material by restricting the access of certain readers. Similarly, they help to prevent certain operators and users from trying procedures or features that they have not been cleared (or taught) to use. Occasionally, it is even cheaper to have several versions. Sometimes we need hundreds or thousands of copies of a short publication but only a few copies of the longer one.

Usually, though, individualized documentation is more expensive, and it can be extremely difficult to maintain. Obviously, it's hard enough to keep one manual current, let alone several versions of it.

Individualized versions can also sometimes underestimate the abilities of audience members, prevent them from learning skills that would increase their value, or even force them to consult several documents.

Most documentors, of course, do not analyze their documentation needs in this way. They think of user documentation as one entity, one file of literature and data. Sometimes this encyclopedic manual is the right choice; for simple systems with a homogeneous set of users, a single manual may be best. But usually the single manual is an expedient choice, a way of simplifying the documentor's planning and production and reducing short-term cost, without much regard for its usefulness to the readers. Furthermore, many of the firms that purvey encyclopedic documents neglect to include an index. An encyclopedia without an index and a system of cross-references is nearly impossible for readers to use. Yet, it still does not automatically follow that the more manuals the better.

Interestingly, a good way to deliver a huge, encyclopedic manual is as an online book, with a utility for looking up key words and phrases.

This way, the difficult search routines, the cross-references and detours, are borne by the system instead of the user.

The problem, of course, is that pages written for paper—especially for an old-fashioned, densely printed, paragraph-filled manual—look terrible on the computer screen. Good online manuals need to be written for the screen.

Exhibit 6.8 Trade-offs between Encyclopedic and Individualized Manuals

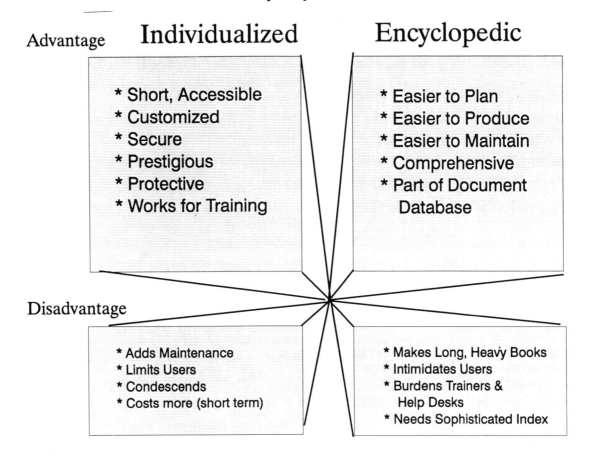

Advantage

Individualized

* Short, Accessible
* Customized
* Secure
* Prestigious
* Protective
* Works for Training

Encyclopedic

* Easier to Plan
* Easier to Produce
* Easier to Maintain
* Comprehensive
* Part of Document Database

Disadvantage

* Adds Maintenance
* Limits Users
* Condescends
* Costs more (short term)

* Makes Long, Heavy Books
* Intimidates Users
* Burdens Trainers & Help Desks
* Needs Sophisticated Index

7. DEVELOPING A MODULAR OUTLINE

7.1 Conventional Outlines: Functions and Flaws

Conventional outlines give very little of the information needed in workplans; they do not specify the length or scale of the sections or the document as a whole; they give no clue to the production costs of the manual. As tables of contents, moreover, they fail to help readers find what they need to know.

Conventional outlines organize the sequence and hierarchy of a text. To do this, they use a tiered scheme of numbers (or numbers and letters) to show subordination and a set of topic headings to show the content or meaning of each section in the document. The typical heading contains only nouns and modifiers.

These conventional outlines are the single most common and useful tool in planning all documents. But are they really powerful enough to perform all the design functions needed in developing structured, modular publications? Their principal benefit is helping writers to organize their own thoughts. They are the perfect planning tool for the artist working alone!

But what about other functions? Can a conventional outline help the designer of a manual estimate its length or the resources needed to prepare it? Does a conventional outline provide meaningful instructions to the several authors who must write the text? Does it generate a useful table of contents?

When a writer works alone on a relatively small assignment, the conventional outline is often an adequate plan. But when teams of writers work on complicated manuals, the conventional outline is not enough. It does not tell the individual authors and contributors how much to write. Neither the numbering scheme nor the typical way of writing headings (without verbs, verbals, or thematic language) tells the writer

how long the sections should be or what they must cover.

The manager or analyst responsible for the publication gets very little data from a conventional outline. Nowhere mentioned are length or the number and type of graphics—often the most important predictors of cost and production headaches.

Ultimately, this uncommunicative workplan becomes an uncommunicative table of contents. And, to the extent that the reader must perceive the hierarchy in the outline, the two-dimensional outline will be useless in guiding the reader through what is really a one-dimensional product!

This last point, obviously, is a bit esoteric and needs explanation. When I say that books are one-dimensional, I am talking not about the way they are conceived but the way they are processed by readers. Although one paragraph may be logically subordinated to another, in reality it is read *after* the other. There is no actual hierarchy in a book, or even in a series of screens to be read; there is a sequence. Item 2 is not really below, beside, or behind Item 1; it is *after* it.

Human readers most resemble computers in that they read in sequence (not in parallel). But human readers are far less adept in assembling a sequence from a maze of loops, detours, and GOTOs.

Exhibit 7.1: Typical Outlines Cannot Answer Key Questions

Management Questions

How Long? What Cost?
How Many People? Days?

Writer's Questions

How Much Writing?
What to Emphasize?

Reader's Questions

Where to Read?
What Matters Most?

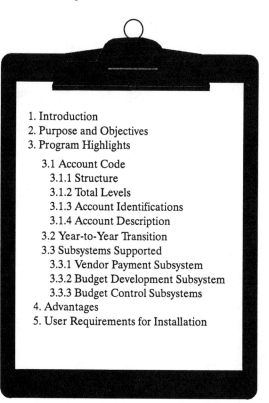

1. Introduction
2. Purpose and Objectives
3. Program Highlights

   3.1 Account Code
     3.1.1 Structure
     3.1.2 Total Levels
     3.1.3 Account Identifications
     3.1.4 Account Description
   3.2 Year-to-Year Transition
   3.3 Subsystems Supported
     3.3.1 Vendor Payment Subsystem
     3.3.2 Budget Development Subsystem
     3.3.3 Budget Control Subsystems
4. Advantages
5. User Requirements for Installation

7.2 Requirements for a Modular Outline

Everyone who plans a document will start with a conventional outline, which is converted in stages to a structured, modular outline in which each heading corresponds to one module of standard size and layout, and in which the language of the heading is informative and thematic—making it more a headline.

Most people do not have thoughts that fit into standard-sized parcels. Thus, their outlines reflect ideas that vary in length and complexity. There is no reason to believe that item 2.1 in an outline will define a section equal in length to that defined by item 2.2. And there is no way to know whether 2.2 is longer or shorter than 2.2.1.

The headings in the outline are little help. Just knowing that item 2.1 is called Administrative Aids and 2.1.1 is called Access Subsystem does not solve the problem.

For an outline to serve the design functions needed in a structured approach to user documentation, it needs two things that most conventional outlines lack:

- a style of language that specifies for the writer, reviewer, and—eventually—the reader what exactly is covered in each section
- a standard that requires each entry in the outline to correspond to a certain standard-sized "chunk" of material

Now, of these two requirements, the first is far less exotic than the second. Many skillful writers use headlines rather than traditional headings—if not in their original outlines then in their tables of contents. For decades, many technical writers have avoided the traditional "Account Code Assignment" (three nouns) in favor of "Assigning the Account Codes" or "How to Assign Account Codes" or "Six Rules for Assigning Account Codes" or even "Why You Won't Need to Assign Account Codes."

The other suggestion—that each entry in the outline correspond to a standard-sized item of material—is less familiar to most writers, unless they are experienced with defense and aerospace technical proposals, in which the technique is commonplace. In fact, many analysts and technical writers are astonished at the idea. How can it be possible to arrange the ebb and flow of ideas into units of uniform size? Is it feasible? Is it worth the effort?

Every communication is necessarily organized into standard-sized units already. Most notably, manuals are already organized into pages of uniform size; no matter how free-flowing the ideas, they are packed in one-page chunks. Most writers and editors, though, leave this packaging up to chance. They rarely know how many pages an idea will take; they cannot predict the length of their discussions. In effect, they let the people and machines that process documents decide where pages will break.

In contrast, in the structured approach to documentation, the goal is to design the actual object the reader/user will see: the pages or screens that will be read. If the book is naturally organized by pages, why not plan and design it page-by-page? If the idea of making every section or unit the same size seems impossible, then why not make them all about the same size, with an upper limit that all must meet?

Why not convert the conventional outline to a list of specifications for each of the modules in the emerging manual or information product?

Exhibit 7.2: Differences between Conventional and Modular Outlines

In Conventional Outlines	In Modular Outlines
■ The entries are cryptic, clear only to the author.	■ The entries are substantive and informative, clear enough for review and testing.
■ The entries correspond to no particular length or size of document.	■ Each entry corresponds to a standard physical entity, of known length.
■ The conversion to a physical product is left to editing and production of the finished draft.	■ The format and layout of the physical product is inherent in the outline (provided the module has been defined).
■ The scope and cost of the proposed publication are NOT apparent from the outline.	■ The outline constitutes a workplan from which costs and production requirements can be estimated easily.

7.3 Defining a Module of Documentation

Modules of documentation may take many forms, as long as each module is small, is cohesive enough to be independent from the other modules, and addresses a single function or theme. Each module should be synoptic, that is, show the entire content of the module in a single array that will not force the user to turn pages. Modules, then, must be on one standard page, or on one odd-sized page, or on one screen-panel, or on two facing pages.

A modular publication is a series of small, cohesive chunks of technical communication of predictable size, content, and appearance. Once the design—the exact sequence—of the modules is frozen, it becomes possible to treat the one, large, complicated manual as a set of small, nearly independent manuals.

Each takes only an hour or two of effort to write; each can be developed independently of the others, in any sequence. With a modular plan, it is even possible to number all the pages and figures as they are written, even though they have been written out of order!

Exhibit 7.3: Tracey's STOP Module

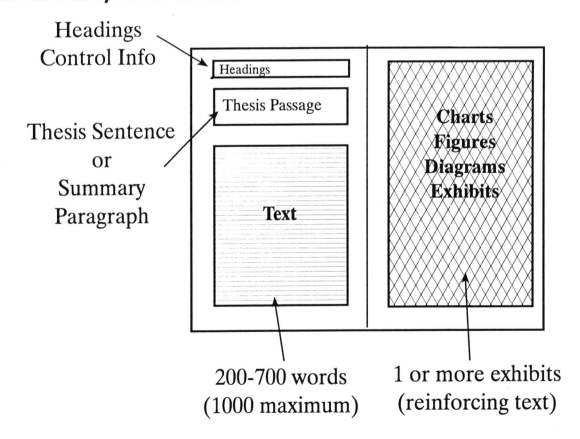

Headings Control Info

Thesis Sentence or Summary Paragraph

Headings

Thesis Passage

Text

Charts Figures Diagrams Exhibits

200-700 words (1000 maximum)

1 or more exhibits (reinforcing text)

The easiest notion of the module is one page. But the one-page limit, although it is perfect for some publications, is troublesome for most. It is too short to present any but the simplest concepts, without crowding the page with tiny, cluttered lettering.

The two-page module made up of two facing pages (a spread) is harder to maintain than the one-page module, but more versatile in use. The best arrangement (developed, as already mentioned, over 20 years ago by the Hughes Aircraft Corporation) is like that shown in Exhibit 7.3.

- **Headings** contain the document number, page, and other typical control data; the name of the section; and the headline (thematic or functional heading).
- The **thesis passage** (summary) lays out the central idea of the module.
- The **full text** (usually 200 to 700 words) expands the ideas in the headline and summary.
- The **exhibits**—screens, diagrams, tables, drawings—are on the right-hand page, or intermingled with the text.

Note that a two-page spread is what people see whenever they open a publication—whether or not the writer intended it. This modular approach merely *unifies the logical document as conceived with the physical document as received.*

A module may fit on one page, even one 11" by 15" piece of computer printout. Or the arrangement of the parts in the two-page module may be changed: some documentors prefer exhibits on the left; some like to shift from right to left for the sake of variety.

Before describing some of these alternatives, however, two ideas must be stressed:

- The definition of the module, though standard and fixed, is really quite flexible, allowing great variability in the number of words or exhibits in the module; especially in the two-page module, the actual "heft" or bulk of the module can vary considerably from one ostensibly same-sized module to the next.
- The particular two-page format recommended here has more than 20 years of successful application in all aspects of technical, industrial, government, and business communications; it is easy to learn and remarkably effective.

7.4 Alternative Forms of the Module, for Special Needs

The recommended two-page module is appropriate and useful for most user documentation. It is synoptic (no page turning) and is large enough to give the advantages of modularity. There are, in addition, one-page alternatives, as well as a variety of specially sized pages and job aids.

The basic module contains one page. Exhibit 7.4a shows alternative arrangements of that one page. Generally, one-page modules are best for orientation/tutorial publications, where the scope of each module is small. Although integrating text and artwork on the same page (rather than facing pages) demands a higher level of word processing or publishing software, for most firms this is no constraint.

Even when the one-page module is technically feasible, though, note that, as in software engineering, the smaller the modules, the more complicated the interfaces: the greater the number of references, loops, skips, and detours. Very small modules that do not allow space to repeat

material force you to send your reader to other modules.

Starting with the presumption that a module will be half text and half exhibits, you may then discover that in many modules it makes sense to change the mix. A module may contain mostly text, although it is unwise to do that too often; more often, it will contain almost entirely charts, listings, or some other exhibit. You may also shift the exhibits to the left, or even expand the module with a foldout.

Exhibit 7.4b shows the possibilities of odd-sized pages: half-size pages, or smaller, for very simple machines; 11" by 15" printout pages; 2' by

Exhibit 7.4a: One-Page Modules

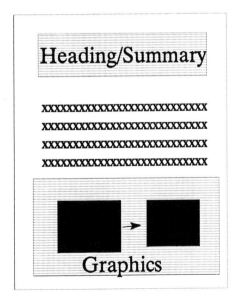

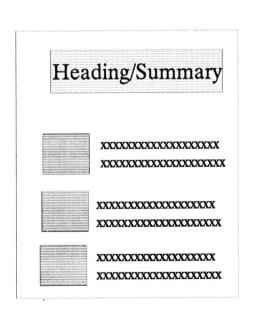

3' posters; fanfold reference cards; keyboard accessories.

A module may be of any workable size or shape, as long as it is big enough to communicate a whole concept, small enough to be easily specified in a plan, and synoptic.

Thus, the perfect module might be a single screen—preferably one that could be read without scrolling. In that case, the size of the module would be constrained by the limits of the video display.

Exhibit 7.4b: Special-Purpose Modules

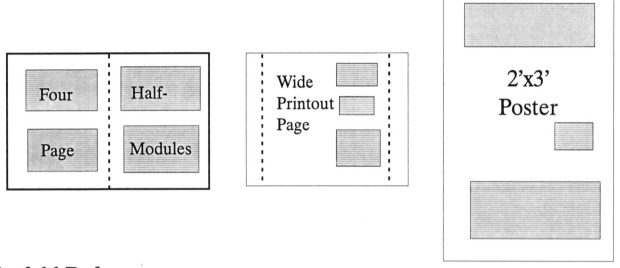

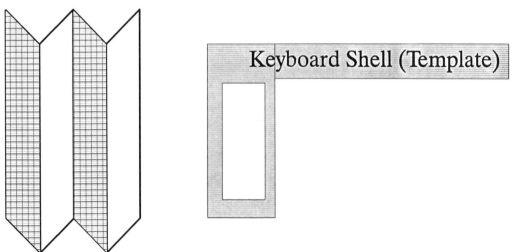

7.5 Writing Headlines for Modules

Each module, regardless of its size and shape, is impelled by a headline. *Headlines, unlike traditional headings, contain themes, ideas, assertions, even arguments. In contrast, traditional headings, even when they are detailed, usually contain only modifiers and nouns.*

The traditional heading used in outlines and tables of contents contains only nouns ("Logon," "Logon Procedure," "Power Redundancy") or nouns and modifiers ("Alternate Logon Procedure," "Multiple Power Source Redundancy").

These traditional headings give little clue to the actual scope or intent of the section and certainly no clue to its length. Neither the author who must write the section nor the reader who is searching the table of contents really knows what the writer of the outline had in mind.

In contrast, **headlines** express ideas, themes, emphases. Both the author and the reader know why the section is written and what it is supposed to do (Exhibit 7.5a).

Exhibit 7.5a: Styles of Headlines

STYLE	EXAMPLE
VERBALS	▪ Backing-Up the Mailing List ▪ How to Import Text Files ▪ Two Ways to Open an Account
CLAUSES	▪ What Happens to Deleted Records ▪ When You MUST Call the Help Desk ▪ Why It Is Safest to Return to the Main Menu
APPOSITIONS	▪ New Accounts: Who May Authorize One? ▪ Security Fence: Which Terminals May Write ▪ CGM: Adding Graphics to the Text
INJUNCTIONS/ THEMES	▪ The Need to Save Frequently ▪ The Importance of Updating Subscriber Lists ▪ The Risks of Unverified Transfers
SENTENCES	▪ This Demo Version Cannot Print ▪ Memorize Your Password ▪ Does Everyone Get the Monthly Statement?

The real key to writing an effective headline is knowing exactly what you want said in the module. Note: *Not* "what you want to say" but "what you want said." Writing good headlines may be the first step toward getting other people to help you write the manuals you do not choose to write alone.

Note also that the knack of writing headlines is, in another sense, independent of the knack of organizing a manual into standard-sized modules. Clearly, one can write headlines without any regard for the heft or length of the material to be covered under them. In fact, one way to develop a modular outline is to go through the intermediate stage of a **substantive outline**, one in which traditional headings have been recast into head-line style without regard for module size. The substantive outline is then refined (disaggregated) into the structured outline.

Although it may seem somewhat confusing to solve both problems at once—thematic language and module-sized chunks—many writers find it easier to plan this way. Knowing the size of the module tends to refine the headline, making it sharper and clearer.

Exhibit 7.5b shows some traditional headings taken from my own library of user manuals and also what their authors seem to have meant by them. (I have left a few blank, just in case you want to try your hand at a hypothetical headline or two.)

Exhibit 7.5b: Converted Headlines

Before	After
Access Methods	Two Ways to Access Files: Sequential & Direct
Files	How the System Validates File Requests
Executive Libraries	The System Includes a Library of Report Templates
Program DEBUG	How to Edit Programs with DEBUG
FDEBUG Example	Debugging a FORTRAN Program with FDEBUG
Transparent Write	_____
Tape Storage	_____
Weekend Testing	_____

7.6 Demonstration: Headings into Headlines

When traditional outlines are converted to substantive outlines, they often look like lists of headlines. The tone of the headlines may be either light and conversational (a kind of "marketing" style) or, alternatively, straightforward and technical.

Exhibit 7.6a shows the outline for an installation plan in both traditional and substantive forms. The "before" version is typical of the style of data processing departments, but the "after" version is vastly more likely to communicate clearly with the user departments and others affected by the installation.

Exhibit 7.6b shows the outline for a typical user guide to a typical accounting package. Note, however, that there are two versions of the revised outline: one showing the appropriate language, scope, and sequence for an executive in the financial department, the other showing a financial clerk how to operate the system.

Exhibit 7.6a: Converting the Outline for an Installation Plan

Before:

1. Site Preparation
    1.1 Electrical Requirements
    1.2 Physical Requirements
2. Assembly
    2.1 Attachments
    2.2 Interfaces
3. Communications
    3.1 Communication Protocols
    3.2 Alternative Configurations
4. Testing
    4.1 Communications Test
    4.2 Mechanical Test
    4.3 Software Test

After:
1. Installing the Necessary Electrical Fixtures
2. Ensuring the Right Temperature and Cleanliness
3. Attaching the Cover and Paper Feeder
4. Choosing and Attaching the Right Cables and Connectors
5. Connecting the Plotter to the Computer
6. Setting the DIP Switches
7. Running the Communications-Check Program
8. Diagnosing Start-Up Problems
9. Solving Communication Problems
10. Solving Mechanical Problems
11. Setting the Switches for Your Graphics Software
12. Testing the System with Your Software

Notice also how traditional outlines do almost nothing to help the writer anticipate the audiences and functions of the publication.

Those decisions are not made until the writer begins the rough draft.

Exhibit 7.6b: Two Outlines for a Business Product

Before:

1. Introduction
2. Accounting Highlights
    2.1 Account Structures and Levels
    2.2 IDs and Descriptions
3. Systems Supported
    3.1 Vendor Payment
    3.2 Budget Development
    3.3 Budget Control
    3.4 Financial Reporting
4. Appendix: Sample Outputs

After: Executive Version

1. Using a Financial Information System
2. Defining Accounting Codes to Meet Legal Requirements
3. Defining Accounting Codes to Support Planning and Analysis
4. Customizing Financial Reports
5. Analyzing Current Patterns of Expenditure
6. Simulating Alternative Budgets
7. Enforcing a Budget

After: Clerical Version

1. Entering Data
    1.1 Entering a Receivable
    1.2 Entering a Receipt
    1.3 Entering a Payable
    1.4 Entering a Payment
    1.5 Entering a Budget
    1.6 Editing a Mistake
2. Getting Reports
    2.1 Running the Monthly Report
    2.2 Running the Quarterly Report
    2.3 Running the Year-End Report
    2.4 Running the Annual P&L Statement
    2.5 Running the Budget Comparisons
3. Appendix: How to Respond to Error Messages

7.7 How Outlines Develop

Modules develop through successive approximations, each with added layers of detail. Instead of the traditional form of writing—from vague topical outline to first draft in one move—we see a series of increasingly richer outlines: topical, substantive (rich in language), modular (one entry per module).

Typically, outlines contain mainly nouns, with a few adjectives for refinement. These topical maps are sufficient for small, uncomplicated writing projects, especially if the writer works quickly, before the encrypted meanings of such entries as "Communication Protocols" escape.

For complicated documents though—those in which there is a more than trivial risk of putting the wrong material in the wrong sequence—the traditional outline is an inadequate design tool. Moreover, its opacity to everyone but its author makes this kind of outline unuseful in any prior review of the emerging document.

The first step involves **titling**, the craft of labeling the sections of the document with substantive headings. In this step, technical writers are expected to apply the craft of journalists, replacing headings with headlines. The headlines should answer the question: What themes, ideas, or practices will be addressed in "Communication Protocols"? The answers must come from the designers; only they know what they mean:

Exhibit 7.7a: Stages of Outlining

Traditional Communication Protocols

Substantive Setting the DIP Switches

Modular How to Read DIP Switches

Setting *MODEM* DIP Switches

Setting *PRINTER* DIP Switches

Setting *PLOTTER* DIP Switches

- How to Set DIP Switches
- Linking with the Mainframe
- Are Your Jumpers and Patches Installed Correctly?
- A Fast Way to Overwrite Jumpers with Software
- What Modem and Cable You Will Need

Again, only the designers know for sure.

Once each of the topical entries is translated into a richer thematic, substantive entry (the substantive outline), the next task is to "decompose," or disaggregate, these entries into module-sized chunks.

As Exhibit 7.7b shows, the issue is whether the matter defined in the substantive heading will fit comfortably within the limits of one module (however that has been defined). In some cases, the answer is yes, and that headline slides through unaltered into the modular outline. In more cases, though, the original headline needs several modules to develop its points. When this is so, the expanded set of headlines (one per module) is transcribed into the modular outline.

Note that a procedure which is too big for one module usually needs a **hierarchy** of modules, not just a series. If, for example, the original substantive heading were "Printing a File," then the expanded, modular outline might be

1.0 Printing a File: The Three Stages
   1.1 Retrieving and Verifying the File
   1.2 Setting the Printer/Format Options
   1.3 Distributing the Output

Exhibit 7.7b: The Process of Outlining

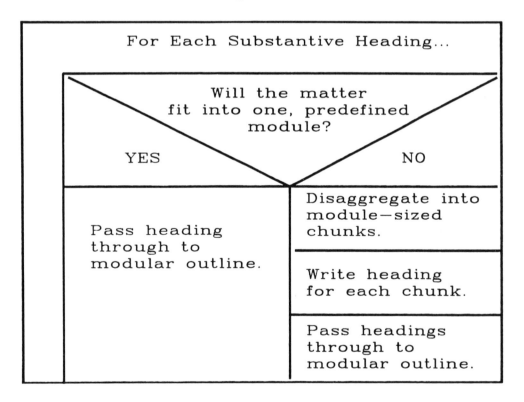

For Each Substantive Heading...

Will the matter fit into one, predefined module?

YES — Pass heading through to modular outline.

NO — Disaggregate into module-sized chunks. / Write heading for each chunk. / Pass headings through to modular outline.

7.8 Option: Reaching the Modular Outline in One Step

Many authors, especially after they have designed two or three modular documents, elect to go directly from the topical outline to the modular. That is, they translate the topical entries directly into thematic headlines for module-sized chunks.

On the surface, it seems logical to reach the modular outline by degrees. First, the strings of nouns and adjectives are translated into the thematic language of the substantive outline. Then each entry in the substantive outline is either carried through into the modular or, alternatively, disaggregated into module-sized chunks.

After a few writing projects, though, many writers find it easier to go *directly* from the topical outline to the modular. Once the size and scope of the standard module become second nature, to them, they are able to write substantive headings that already fit the modular constraint.

Moreover, many writers discover that it becomes easier to invent substantive headings

Exhibit 7.8a: Skipping the Intermediate Conversion

Conventional Outline	Structured Outline
I. Introduction	
1.1 Background	1. Problems in Batch Processing of Retail Transactions
1.2 Project History	2. A Three-Stage Conversion to TRANSACTIONS
II. Operating Highlights	
2.1 Transparency	3. How the New System Simplifies Operation └4. Elimination of Batch Activities
2.2 Transaction-Based	5. Each Transaction Updates <u>All</u> the Files └6. Data Are Typed Only Once
2.3 Security	7. Access Is Strictly Controlled
III. Functions	
3.1 Retail	8. Recording Retail Transactions └9. Opening Retail Accounts
3.2 Accounting	10. How the Retail Data Reach the Accounting Files
3.3 Inventory Control	11. How the Retail Data Reach the Inventory System

when the scope of the material has been defined. They find that they need a sense of scope when trying to choose the most precise and evocative headline for the substantive outline. For example, if the original topical heading is "Envelopes," it is difficult to know the appropriate substantive heading:

- Printing Envelopes
- Printing Envelopes on the Laser Printer
- Printing Envelopes on Laser Printers with a Central Envelope Feeder
- Customizing the Envelope Printing Program

It is hard to say which one (or combination) of these is appropriate—until one decides to outline in module-sized chunks.

Exhibit 7.8a shows a direct conversion from the topical to the modular outline (without the intervening substantive step).

Exhibit 7.8b shows that, sometimes, the order of the original outline needs adjustment. The revised sequence is the planner's design for a more accessible version of the material implied in the topical outline.

Exhibit 7.8b: Altering the Original Sequence

Original Sequence	Revised Sequence
1. Problems in Batch Processing of Retail Transactions	1. How the New System Simplifies... (3)
2. A Three-Stage Conversion to TRANSACTIONS	2. Data Are Typed Only Once (6)
3. How the New System Simplifies Operation	3. Each Transaction Updates... (5)
4. Elimination of Batch Activities	4. Elimination of Batch Activities (4)
5. Each Transaction Updates *All* the Files	5. Opening Retail Accounts (9)
6. Data Are Typed Only Once	6. Recording Retail Transactions (8)
7. Access Is Strictly Controlled	7. How the Retail...Accounting Files (10)
8. Recording Retail Transactions	8. How the Retail...Inventory System (11)
9. Opening Retail Accounts	9. Access Is Strictly Controlled (7)
10. How the Retail Data Reach the Accounting Files	10. Problems in Batch Processing... (1)
11. How the Retail Data Reach the Inventory System	11. A Three-Stage Conversion... (2)

7.9 Is It Possible to Predict the Number of Modules?

People who do not write much are skeptical of the claim that, at this early stage, it is possible to estimate the size of the modules, to be sure that what is defined as a module will fit into the space provided. Actually, within a month most people to learn to estimate the heft of the modules and, thus, the number.

Even professional technical writers, who often make rough estimates of the length of a publication by looking at the outline, are reluctant to predict length precisely. Obviously, it is impossible to tell from a traditional outline the precise length of any of its entries. Fortunately, though, devising a structured outline does not require such an estimate.

In this scheme, a module is an upper limit (one or two pages, of any dimensions), not a uniform size or length. *It is far easier to make sure that no module is too large than to make sure that all modules are equally long.*

Modules can vary considerably in length and content, as long as they all fit into the same two-page spreads. By adjustments of artwork and typography, a module might have as few as 200 words, with or without an attendant increase in the size or number of exhibits, or as many as 1000 words.

Furthermore, the modular outline is not the last chance to estimate the size of the modules. Later, when the outline is finished, the designers will write a small spec for each module, at which point they may decide that what they thought was one module is, in fact, more than one. And, still later, when all the specs are mounted in a story-board, there is one more chance to revise the estimate.

Generally, it takes writers only a few weeks to develop a sense of the module-sized chunk of material. Amateur writers often learn the technique quicker than professionals, who need a few days to unlearn some of their old habits. Although the idea of setting a standard physical limit on the size of a concept or procedure seems harsh and restrictive at first, in a short time it reveals itself to be a useful discipline that encourages intellectual creativity.

After a few years of writing in this modular style—and of encouraging others to learn it—many writers may discover that what seemed procrustean and inhibiting at first is actually liberating and exhilarating. Designing modular publications, like structured methods in general, converts overwhelming assignments into manageable projects. And the discipline of disaggregating complex ideas into module-sized chunks converts overwhelming concepts into manageable bits of comprehensible information.

The two-page spread is, of course, arbitrary. (Someone has said that a standard is an arbitrary solution to a recurring problem.) Or is it arbitrary? Is it not possible that, since books naturally present themselves in two-page spreads, that educated people learn to apprehend intellectual material in just such chunks? Even if there is nothing biological or metaphysical about the two-page spread, cannot we argue that it is a key element in Western culture? Why is it that when people begin to design publications this way they often speak of being "converted" to the method?

Even publications managers, who start out skeptical, are usually won over—unless they are preoccupied with conserving paper.

The only change in publication policy needed to implement modular publications successfully is a willingness to allow some white space in manuals—the consequence of occasional short modules. For some, though, the sight of blank white paper is anathema. They see expense and waste; they do not see the increased readability and maintainability of the publication—which sometimes save thousands of times as much money as the "less wasteful" printing could have saved. To the question "Won't there be a lot of blank space in a modular publication?" the answer is "Probably."

8. DEVELOPING A STORYBOARD

8.1 The Value of Models in Solving Documentation Problems

Models save money and effort. They allow you to experiment and innovate with smaller risk and slight expense. Without models, you are unlikely to test documents as hard as you should.

Exhibit 8.1 depicts one of the most important functions in the world of work: the relationship between the cost of correcting an error and the time at which the correction is made. The function is exponential; that is, the curve not only accelerates, it accelerates at an ever faster rate.

The more complicated the project, or the more unfamiliar and risky the technology, the faster the curve accelerates.

A **model** is a representation of one thing by another. Models are made either from different materials (clay instead of steel, paper instead of switches) or on a different scale (a miniature of a building, an oversized model of an atom). The materials and scale of the model make it easier to build and, more important, easier to change.

Models are relevant to writers of user documentation in two important ways. First, user documents *can be models themselves*. In the most sophisticated development groups, the design team will write an operations guide as a way of specifying and testing the user interface. In other

Exhibit 8.1: Costs of Change over Time

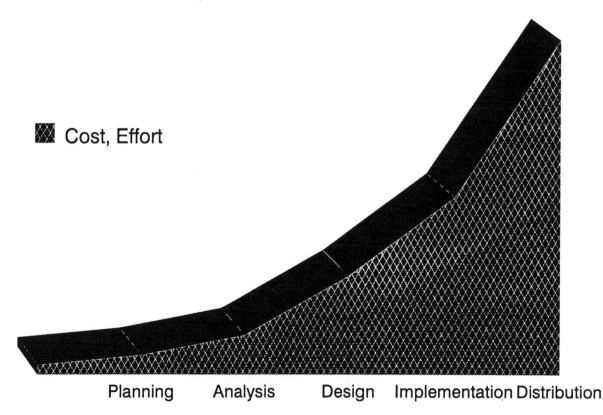

Cost, Effort

Planning   Analysis   Design   Implementation Distribution

words, the developers, by writing a before-the-fact user publication, guide the subsequent design of the system or program. And if that model publication has been written by an expert on efficient, friendly operations, the resultant product will also have those characteristics.

(User documentation written at the beginning of the life cycle is still exceptional, but clearly—as many have discovered—is the best way to work. Among other benefits, writing early user documents forces developers to think about the thing they usually leave for last: what the *people* will do with the system.)

The second relevance of models is in the *development of the documents* themselves. Publications also need models. And the models they need are much more ambitious than a traditional topical outline. Even the substantive and structured outlines described earlier, although they are a vast improvement on traditional designs, are not enough.

Rather, before writing and drawing a draft of any publication longer than a few pages, documentors should devise a model of the publication that makes clear what will happen *within* each module and also that shows *all the links and couplings across modules*. In effect, it should be possible to evaluate the accuracy of the technical content in each module, and predict the number of loops and branches across modules. As in software engineering, *the greater the number of possible paths through the document, the less reliable and more error-prone the process of reading it.*

Models are for testing. And the purpose of testing is to find flaws, mistakes, and bugs. Models and tests encourage you to change your mind, raise or lower your sights. Models and tests make misunderstandings vivid, focus disagreements, underscore schedule and cost problems, and prove that you cannot have everything you want—or cannot have it in time. In short, *models force you to recognize your errors and redo your work.*

And that is why few writers, and nearly as few programmers, want to use them. Most of the clients I meet do *not* want to know what is wrong with their work. They do *not* want to be reviewed, tested, inspected, verified, validated, evaluated, or "walked through."

Obviously, no one likes criticism. But the longer people work on a manual or system, the less receptive they are to critical opinions. An added benefit of working with models is that they enable people to see the flaws in their plans early—before they have fallen in love with them!

8.2 Writing a Specification for Each Module

*For each module defined in the modular outline, the designers prepare a **module spec**, a brief but informative description of the content of the module. The main elements in this spec are a thesis passage (summary) and a sketch of the exhibits that are to appear in the module.*

Given an approved modular outline, the designers write a specification for each proposed module. The **module spec** contains

1. The **heading**—transcribed from the modular outline

2. **Context** (optional), the location of the particular module in some larger hierarchy. In the example below, for instance, the module "How to Add a Node" is part of a hierarchy:
Changing the Configuration (sup)
   How to Add A Node
        Updating the Security Profile (sub)
        Updating the Reports Route (sub)
   This part of the spec may be eliminated for any module that truly stands alone, that is, one unconnected to any other module. The best document database would be a collection of such *unencumbered* files.

3. The **summary** or **thesis passage**—a précis with one-to-four sentences that distill the main information in the module. *Note*: The summary is **informative**, that is, it contains the same information as the module in shortened form; it does not merely promise what will be discussed.

4. **Exhibits**—sketches or clear specifications for the nonprose part of the module. Although not every module will have exhibits, all are *presumed* to need at least one.

5. **Notes**—as needed. Just in case the headline, summary, and exhibit are not clear enough—although they usually are—the designers may also add a few notes describing what the module will contain. Just enough so that someone who had reviewed the spec would not be surprised by the finished module.

These module specs will be mounted in a gallery for review and revision. Included in the review are those people who will be asked to write the missing part of the modules (and who, therefore, are consulted on the overall design of the publication), as well as potential users/readers of the publication.

After a brief "learning curve" (2 or 3 hours), most people find that they can write a module spec in 10 to 15 minutes. And in some publications, in which a series of modules falls into the same pattern, the repetitive module specs can be prepared in only about 5 minutes each.

Exhibit 8.2: Filled-in Module Spec

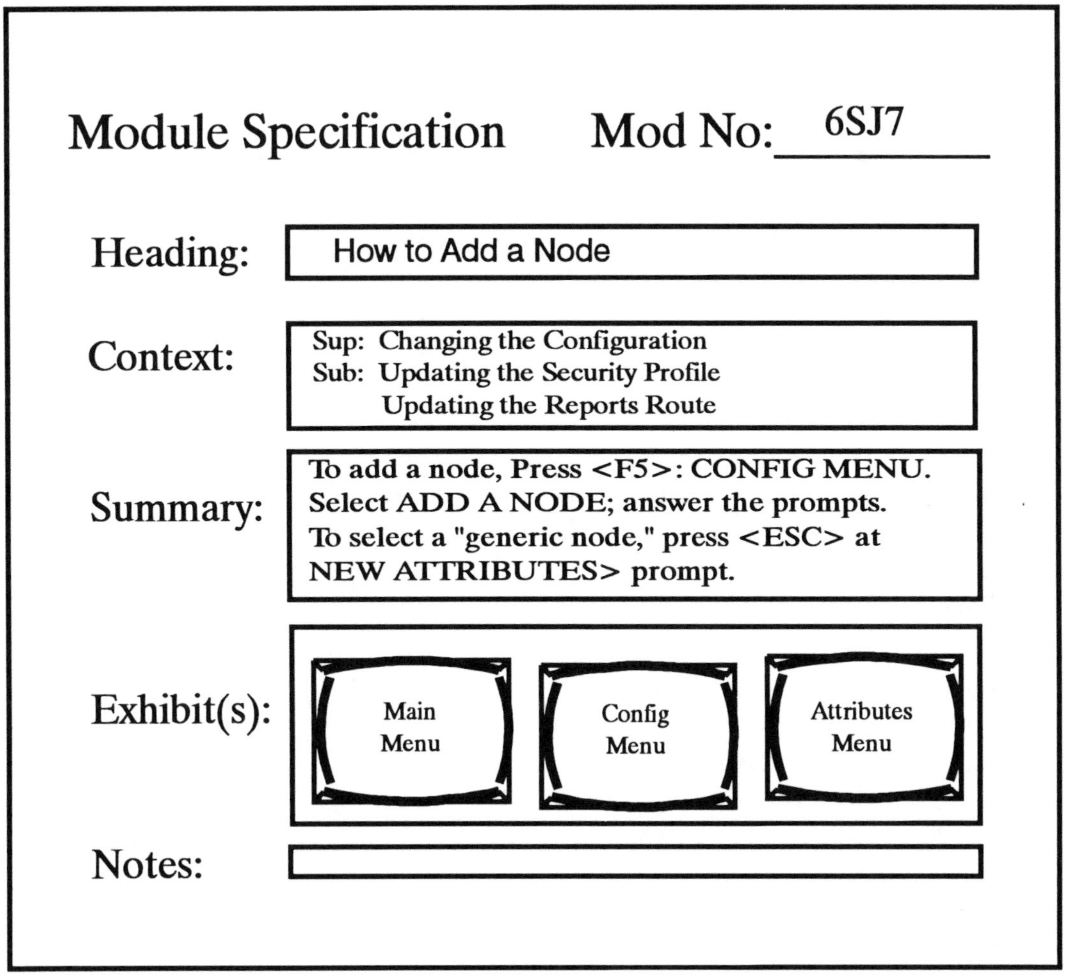

Module Specification　　Mod No: __6SJ7__

Heading: | How to Add a Node |

Context:
Sup: Changing the Configuration
Sub: Updating the Security Profile
　　　Updating the Reports Route

Summary:
To add a node, Press <F5>: CONFIG MENU.
Select ADD A NODE; answer the prompts.
To select a "generic node," press <ESC> at
NEW ATTRIBUTES> prompt.

Exhibit(s):

Main Menu　　Config Menu　　Attributes Menu

Notes:

8.2.1 Does Every Module Need an Exhibit?

Nearly every module in a user manual can benefit from an exhibit—a diagram, a screen or two, a drawing, a word chart. In a well-designed module the exhibit is redundant with the text, not supplemental to it.

Presume that every module will have an exhibit. That is, plan on having at least one exhibit in each module, but be prepared to abandon the idea if, after hard thought, you cannot think of one. Or if there is not enough space.

The material communicated in the exhibit overlaps with and reinforces—in some cases duplicates—the material in the text. In fact, in the best module there is double repetition; the headline and summary state the content, which is echoed in the exhibit, which is further echoed and enhanced in the detailed text. Does not all this redundancy violate some principle of concise, technical communication?

No. Redundancy violates a principle of *economy*; redundancy raises the short-term costs. Indeed, leaving out the graphics altogether also reduces short-term costs. Remember that *redundancy is absolutely necessary to ensure effective communication*. And redundancy of pictures and text is the shrewdest way to communicate technical information to audiences with different learning styles.

Most of the exhibits will fall into these main categories:

- **Displays and screens**—duplications, reproductions, or renderings of what actually appears on the video display or other input/output device; one or more screens per module is the most typical method of documenting online systems.
- **Flow and process diagrams**—abstract symbols that represent either the physical movement of events and material or the logical movement of data and ideas. There are also diagrams that clarify procedures.
- **Drawings and representations**—any attempt by art or photography to depict actual objects or, occasionally, people; technical drawings are usually the preferred method because they are easier to reproduce, but photographs are used when the emphasis is on credibility.
- **Verbal graphics**—exhibits made up mainly of words, with some simple embellishments such as boxes and arrows; although rare in technical or user manuals, verbal graphics can be especially useful in plans, briefings, and training materials. (The module you are reading now has a verbal graphic.)
- **Playscript/dialogue**—techniques that show operators, users, and equipment as though they were following stage directions; playscript, a set of techniques developed originally for manual systems and procedures, is extremely adaptable for data entry.
- **Mathematical and statistical exhibits**—equations, formulas, graphs, statistical tables, and the full range of exhibits associated with science and engineering.

Can the same exhibit appear in more than one module? Of course. Although many technical editors and publications managers will resist the suggestion, I urge you to repeat an exhibit rather than commit that most serious error: referring to an exhibit that cannot be seen.

In practice, however, most documentors discover that each of the several references to the "same" exhibit are, in fact, references to *different* fields on the screen, *different* cells in the table. Although the usual practice is to produce the exhibit once and refer to it from several places in the text, the smarter policy is to create *separate exhibits for each instance*, or, in some cases, the "same" exhibit with different parts emphasized or highlighted.

Exhibit 8.2.1: Types of Exhibits

DISPLAYS	■ Screens/Panels/Windows ■ Worksheets/Forms ■ Messages/Boxes
DIAGRAMS	■ Flowcharts ■ Networks ■ Data Flow Diagrams ■ Structure Charts
PICTURES	■ Illustrations ■ Photographs ■ Design Graphics (Drawings)
VERBALS	■ Word Tables ■ Pseudocode or "Structured English" ■ Decision Tables/Trees ■ "Information Maps" ■ Listings, Programs ■ Playscripts
MATHEMATICS	■ Statistical Plots ■ Pie/Bar/Line/Surface Charts ■ Equations, Models

8.2.2 What If the Material Won't Fit Into One Module?

If the material under one headline is really too big for one module, then it must be exploded or disaggregated. Generally, if an idea or procedure is too big for one module, it needs at least three.

When documentors first try to write in standard-sized modules, one of their concerns is the problem of the module that is just a little too big to fit in the one- or two-page limit. Obviously, if the concept or content is much too large it will need to be treated as several modules. But what of the item that is only slightly overweight?

For those modules that are bursting at the seams, there are numerous remedies. Artwork can be shrunk; even text can be reduced somewhat, although most publication professionals prefer not to change type sizes from page to page.

There are also many ways to expand the capacity of a module without producing clutter and without making it harder to read. Text presented in **columns** usually allows the writer to fit 10 to 20 percent more material into a space without deleterious effects. Text typed with **proportional printing** or **kerning** yields a similar benefit.

There is even the option of *removing* some material from the fat module, provided one is sure that this loss will not interfere with the clarity and effectiveness of the module.

But what of the case that resists these simple adjustments? What of the module that really is too big?

As it turns out, the discovery that a process or concept is too large for one module of documentation is a powerful piece of test data. In almost every case, it means that the process or concept is *too big to be regarded as one entity*. Especially when the module is the spacious expanse of the two-page spread of 8½" by 11" pages, an entity that will not fit the module is probably best regarded not as one thing but as a small collection of things. For that reason, the most common way of redesigning the big module is *not* to break it into two modules, but, rather, into at least three!

As Exhibit 8.2.2 shows, merely breaking a long idea into Phase 1 and Phase 2 is less coherent and intelligible than beginning with an executive view that explains how the process has two phases—and then offering a module for each phase. Thus, a process with two phases needs three modules, with three phases four, and so forth.

In some outlines, the headlines are so broad that it takes a three-level hierarchy to present what was first thought of as only one module. Note, therefore, the advantage of this method of design: Even if the designers have underestimated badly the space needed for many of the modules, the modeling activity will correct the problem.

After a few months of writing in the modular way, writers realize that the failure of an idea to fit into one module is evidence of the difficulty of the idea. *Procedures that can be apprehended and presented within a single module are more easily learned and followed than those needing hierarchies and branches.* Often, then, the wisest thing for a documentor to do—having discovered an especially fat module—is to persuade the developer to change the procedure itself, making it more usable and reliable.

Exhibit 8.2.2: Disaggregating a "Fat Module"

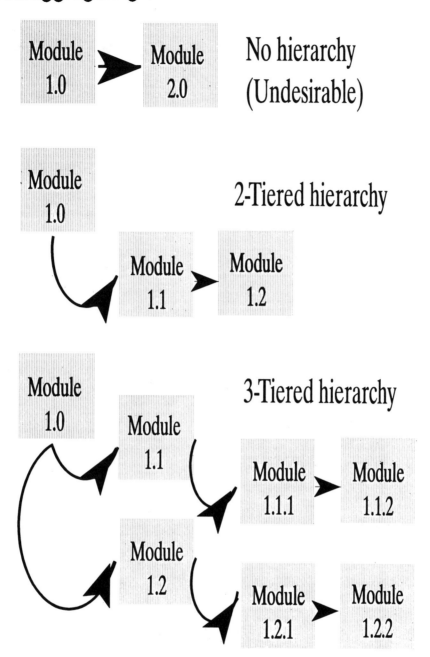

No hierarchy
(Undesirable)

2-Tiered hierarchy

3-Tiered hierarchy

8.3 Modules Must Be Functionally Cohesive

A well-made module addresses a single theme and performs a single function. It motivates the reluctant, orients the novice, guides the competent, or gives reference to the skilled. Each module should limit itself to only one of these functions.

It is not enough to know the subject of a module; one must also know its *function.*

To say that a module "explains the icon menu" is not enough. The question still to be answered is "What function or service will it provide for the people who read it."

Nearly every piece of a user document—nearly every Help screen—addresses some combination of **four functions**. For the brand new, reluctant user, it **motivates**, offers inducements to try things that are strange—even threatening. Once the initial shyness has passed, user documentation **orients** the newcomer, giving elemental definitions and instructions. Eventually, user documentation **guides** the experienced user, demonstrating how to string together the elements into tasks and meaningful work. And, finally, documentation gives the skilled user **reference**, quick reminders of facts that have been forgotten (or never memorized in the first place).

*A well-made module or Help screen will address only **one** of these four functions.* Each of these functions calls for very different communication styles and formats; it is unlikely that a single page can both teach and motivate, and it is even more unlikely that a Help screen which tries to teach procedures and offer quick reference at the same time will be judged helpful.

Even though one publication may address many audiences and purposes, a good module is still functionally cohesive. It addresses not only one topic but one function. Attempting two or more functions in a single module produces a confusing mess of information. More important, though, is the fact that readers typically *need only one of the four functions* whenever they read a particular page. (At this writing, confusion of function seems to be the second greatest flaw in amateur Help screens, the first being unreadable prose.)

Although this issue may seem somewhat theoretical, in fact it explains many of the practical failures in user documentation. Three problems are common:

- Technically oriented writers often ignore the issue of function and, instead, write dull and irrelevant "explanations" or "descriptions" of system features and components.
- Expert writers often provide *reference* alone (what they as experts use) and overlook the other functions.
- To save space and time, well-intentioned writers cram all four types of support into compact spaces, and then are disappointed when users do not read their manuals.

Again, each function needs a different style of writing and different classes of exhibits. Each communicates to a different reader expectation.

Exhibit 8.3: Four Functions of Modules

Motivation
- Inducing reluctant users to try
- Converting features into benefits
- Comparing the new with the old

Orientation
- Preparing the neophyte
- Teaching elementals
- Explaining one thing at a time

Guidance
- Stringing elements together
- Demonstrating whole processes
- Promoting productivity

Reference
- Extending the user's memory
- Answering frequently asked questions
- Enhancing the user's efficiency

8.3.1 Designing a Module That Motivates

A motivational module is one whose purpose is to get the readers to do something they do not want to do. It must convince the readers that they will benefit from the process or technique recommended in the module, that they will gain more from doing what is proposed than from not doing it.

Even though documentors may think of themselves as "technical people," they nevertheless must **sell** ideas and methods to their readers. Operator manuals and user manuals almost always contain some motivational material. That is, modules that convert the features of the system into benefits for the reader.

Every system replaces some other system; the differences between the former and the latter are the features to be described.

Most features fit into relatively few categories:

- **Physical aspects**—components, size, weight, temperature, location, quantities, general appearance, sound
- **Operating aspects**—speed, cycle rate, number of steps, "capabilities" (what it will or will not do), compatibility with other things
- **Accessibility**—quantities on hand, learning time, delivery time, service time, acquisition costs, operating costs
- **Performance features**—elegance, rigor, accuracy, precision, reliability, versatility, expandability

To repeat, any system or procedure you recommend must differ in some of these characteristics from the one you wish to supplant or replace. And the problem is to map one or more of these features onto one or more of the benefits.

The most common mistake is the **features trap**. Many writers think there are a great many

people who find several of the features above inherently desirable and worthwhile. There are fewer of these people, however, than engineers and analysts believe.

As Exhibit 8.3.1 shows, features are converted into benefits, like those proposed in the work of sociologist Harold Lasswell.

Power, for example, is attractive to the executive who wants more control over his or her organization; but it is also attractive to the clerk who wants "free time."

Wealth is the most direct business motivator: the promise that the plan or the product will earn or save money. (One of the hardest sells, of course, is to convince people that high short-term costs will be repaid with higher long-term savings.)

Motivational documents may also appeal to the readers' desire for superiority (**respect**) or for a state-of-the-art challenge (**skill**). Some users are attracted to ease-of-use and reduced stress (**well-being**), while others want to do what is popular with their group (**affection**).

Less often used in business and government is the appeal to **rectitude**: doing something because it is right, or just, or ennobling. And still less often the appeal to act in a way that enhances knowledge and wisdom (**enlightenment**). In some specialized institutions (like universities or religious organizations) and even in some entire cultures, these appeals persuade people to try new activities as eagerly as most people in our society are drawn to the "better bottom line."

The point to be stressed is that none of these benefits is obvious in the features. Even the cost features of a new system may need extensive explanation and justification to prove that they provide a material benefit to the reader.

The documentor must analyze what the readers/users want and must show explicitly—in the summary paragraph of the module—how the recommended action can get it for them. And the exhibit should, in most cases, show the comparative advantages of the two approaches side-by-side.

Exhibit 8.3.1: Converting Features into Benefits

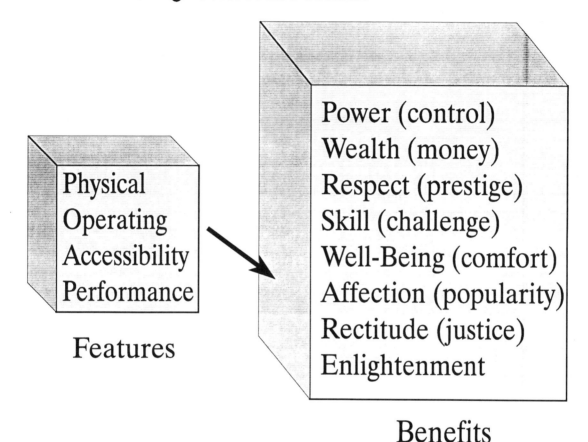

Physical
Operating
Accessibility
Performance

Features

Power (control)
Wealth (money)
Respect (prestige)
Skill (challenge)
Well-Being (comfort)
Affection (popularity)
Rectitude (justice)
Enlightenment

Benefits

8.3.2 Designing a Module That Orients the Novice

An orientation module *teaches a single concept or task and then tests the reader to see if the concept or task has been learned. Documentors define the aim of the module in terms of a particular item to be mastered and then require the reader to prove mastery: by answering a question, completing a simple operation, or advancing along a progression of tutorial instructions.*

An **orientation module** contains *one new thing.*

Before writers can design such a module, they must be able to say exactly what they want the reader to learn from it. And the most useful way to describe that objective is to think of some task or test, keyed exactly to the concept or idea being taught. In effect, if the reader can answer a certain question, make a certain choice, finish a certain process, or otherwise prove mastery of the concept, then the module will have been effective. In more-sophisticated teaching materials, one may even specify other limiting conditions, such as how much time is allowed for the task, or how many wrong attempts are permitted among the right answers.

The sample in Exhibit 8.3.2a might look painfully obvious to an experienced operator, but it frequently is just the right way to communicate with a novice. (Note: Orientation modules frequently take very little room; it is not uncommon to present them in one page, or even in pages that are smaller than the conventional 8½" by 11".)

The sample in Exhibit 8.3.2b is more typical of operators' materials. Naturally, since the task is to generate the "solution screen," the manual must be used at a live terminal or PC. (It is difficult to imagine a way to present a series of such modules without having the reader at a working system. Especially for the inexperienced reader, it is nearly impossible to learn basic tasks without actually doing them.)

Exhibit 8.3.2a: Basic Question

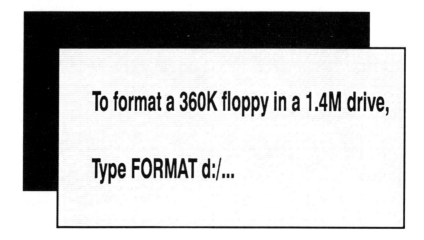

To format a 360K floppy in a 1.4M drive,

Type FORMAT d:/...

Exhibit 8.3.2b: Simple Task

Using the INSERT and CURSOR keys
Change Screen 1 to Screen 2...

WKS	**WORKSHEET**

Exhibit 8.3.2c: Multiple-Choice Question

Another name for "flexible disk" is:

A. Winchester disk
B. Floppy disk
C. CD-ROM
D. Slipped disk

The sample multiple-choice question in Exhibit 8.3.2c could appear either in a book (probably a programmed textbook) or, better, as part of a computer-assisted series of instructional screens. Interestingly, although a programmed textbook is probably the most effective way to teach a novice user, the best programmed texts are designed to force the reader to skip, jump, branch, and detour! The problem is that insecure or inexperienced readers who get lost in a programmed text may never find their way back.

The solution is the online tutorial, a programmed text with branches invisible to the user.

Obviously, more-complicated training materials call for the skills of a specialist, an instructional technologist.

8.3.3 Designing a Module That Guides the Experienced User

Unlike the orientation module, which teaches one small item of information to a novice reader, the guidance module *teaches one whole function, task, or activity. It must be simple and clear—and above all accurate.*

As long as they are clearly written and uncluttered, **guidance modules** can present substantial chunks of information: complete procedures or transactions, whole programs or modules of programs.

The reader of such a module expects it to be accurate. That is, if the procedure in the module is imitated, the result should be as promised. If that is not so, the reader blames the writer of the documentation and the developer of the system. (This contrasts with novice users, who tend to blame themselves.)

As Exhibit 8.3.3 shows, the first task in planning a demonstration module, or a hierarchical series of them, is to be sure the intended reader is an experienced, confident learner, free from the special needs discussed earlier in connection with orientation modules.

The next task is for the designer of the module to write a summary of the process to be described—usually how the person is supposed to do something. This summary should be terse: a list of instructions and conditional actions, which is then tested for accuracy.

In writing up a procedure that already exists, the test is straightforward: We get someone to follow the instructions (and only the instructions) to see whether the program or device performs as expected. If the system is still under development, however, the summary of the process must be tested by having the reviewer verify its correctness—a procedure not as good as a live test, but the best possible in the circumstances.

The most interesting document design problem, at this point, is to decide whether the process or transaction is a one-level or multi-level procedure. In simple terms: to decide whether or not it will fit into one module.

If it is at one level, if everything that needs to be said about the procedure can be handled in the one-page or two-page module, then the designer writes the thesis passage, sketches the simple procedural diagram, and considers the module specified.

But what if the whole transaction calls for more than one module, as many do?

If a process is too big for one guidance module, then it will need a hierarchy of them. That is, it will need an overview module, followed by a series of modules for each main component of the process (a two-level hierarchy); in other cases it may call for a three- or four-level hierarchy.

Defining the hierarchy or components of the process calls for ingenuity; there are always several ways to break a complicated thing into its components. The best way is the one that lets the manual score highest on the Usability Index, that is, the one that reduces the amount of skipping, branching, and looping.

Exhibit 8.3.3: Designing a Guidance Module

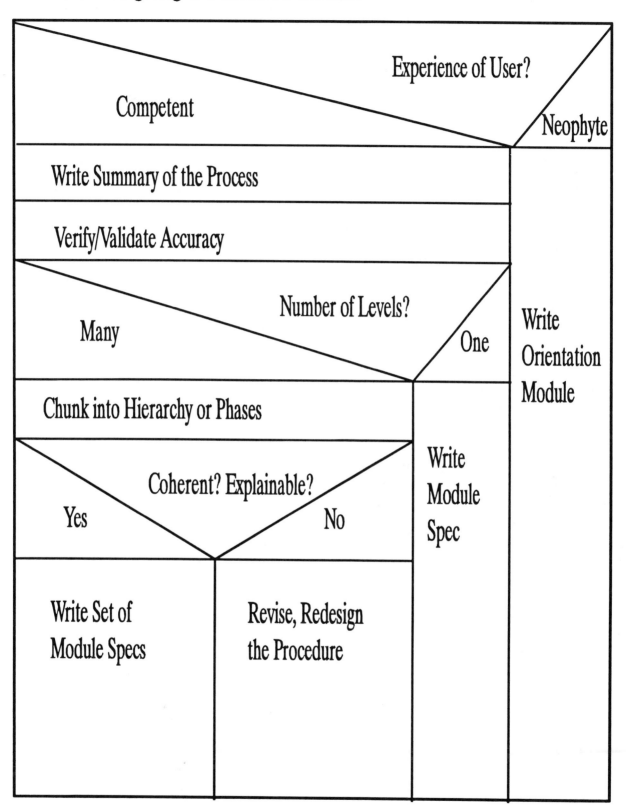

Experience of User?

Competent

Neophyte

Write Summary of the Process

Verify/Validate Accuracy

Number of Levels?

Many

One

Write Orientation Module

Chunk into Hierarchy or Phases

Coherent? Explainable?

Yes

No

Write Module Spec

Write Set of Module Specs

Revise, Redesign the Procedure

8.3.3.1 Replacing Prose with Structured Text

Writers of guides and instructions should favor forms other than the prose paragraph; even educated readers have trouble following directives hidden inside paragraphs. Instead, prose should be replaced with lists, word-tables, scripts, and other structured formats.

Procedure writers should be suspicious of any instructions couched in paragraph form. Even though there are many readers who do read paragraphs, there are many more who cannot— and still more who will not. Business and government readers are skimmers, not studiers; many go weeks at a time without reading a paragraph all the way through.

But even if this were not so, paragraphs would still be suspect. As the Internal Revenue Service knows especially well, even the clearest instructions embedded in paragraph form produce a high incidence of errors and frustrations.

Whenever possible, multi-step procedures should appear in itemized lists or word-tables. Consider the following actual specimen:

BEFORE

The number of days of sick leave which may be credited to an employee, other than a part-time employee, shall be determined by multiplying the total months of service by two and subtracting therefrom the number of days of sick leave previously taken.

AFTER

To compute the number of sick days credited to a full-time employee:

1. Count total number of months of service.
2. Multiply by 2.
3. Subtract the number of days already taken.

OR

Sick leave for full-time employees =

[(# of months service) X 2] – (days already taken)

Although there are some math-phobes who could not handle the second version, everyone would find the first revision preferable to the original paragraph, especially with its nineteenth-century prose style.

Consider also the following typical example:

BEFORE

Persons with two 360K floppy disks should make backup copies of the distribution disks and use the backup of the program disk in Drive A and the backup of the files disk in Drive B. Persons with a hard disk should insert the program disk in Drive A, type **install** and follow the instructions. (To install on one high density, 5.25" floppy, insert the program disk in Drive A and type **install5**; to install on either a 720K or 1.4Meg 3.5" floppy, type **install3**.)

SYSTEM	PROCEDURE
2 360K floppy drives	1. Copy distribution disks. 2. Insert copy of program disk in Drive A 3. Insert copy of files disk in Drive B
hard (fixed) disk	1. Insert program disk in Drive A 2. Type **install** 3. Follow Instructions on screen
1.2M 5.25" floppy	1. Insert program disk in Drive A 2. Type **install5** 3. Follow Instructions on screen
760K 3.25" floppy	1. Insert program disk in Drive A 2. Type **install3** 3. Follow Instructions on screen
1.4M 3.25" floppy	1. Insert program disk in Drive A 2. Type **install3** 3. Follow Instructions on screen

Even in these simple procedures, the tabular form is conspicuously easier to follow than the paragraph. And the benefits are even greater when the instructions are aimed at two or more users. When a procedure passes from person to person, the clearest way to present it is with the technique known as playscript:

BEFORE

To get access to the files of another user on the LAN, you must get the owner of the files to grant written permission, specifying your read/write privileges on Form MIS89-10. This form must be sent to the LAN Administrator who, after receiving the form, has 5 days to create the software links necessary, consistent with the read/write privileges. (For read-only links, the LAN Administrator must respond within 3 days.) Upon receipt of an e-mail bulletin from the Administrator, you may access the designated files.

AFTER

Actor	**Action**
Applicant	1. Tells file owner of access request
Owner	2. Completes form MIS89-10 2a. If denied, advises applicant
LAN Administrator	3. Creates necessary software link 3a. If read/write, within 5 days 3b. If read-only, within 3 days 4. Sends e-mail bulletin to applicant
Applicant	5. Accesses the file, as needed

8.3.3.2 Replacing Prose with Decision Graphics

Whenever a procedure involves decision-making or branching, words should be enhanced with arrows or other logical markers. The best plan is to use tree diagrams and other decision graphics.

Even clearly written instructions become difficult when readers are asked to follow a complicated path. Although techniques like playscript allow for "sidetracks" and other branching operations, it is not an exaggeration to say that the more often the reader must read something other than the next line, the less suitable is ordinary prose.

Countless procedures could be improved by converting prose to special combined forms, decision trees, or decision tables.

The first example uses a Nassi-Shneiderman Chart, a technique developed for structured design of computer programs but also well-suited for "manual" procedures:

BEFORE:

To delete a terminal from the access list, first bring up the list using <PF17>. If the terminal is not currently on the list, do nothing. If it is, press <PF10> (modify list), select the terminal to remove, and press <Enter>. Repeat for each terminal to be removed.

AFTER: (Exhibit 8.3.3.2a)

Exhibit 8.3.3.2a: Converting Prose to Nassi-Shneiderman Chart

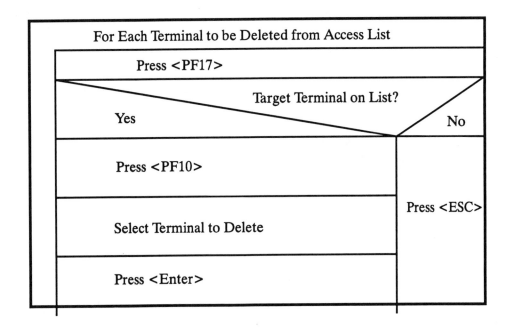

In the next illustration, the text is converted to a decision tree:

BEFORE:

If you receive the "Illegal Access Attempt" message, determine whether you have mistyped the name of the file. (If you have, retype and continue.) If the file name has been typed correctly, review your access privileges by pressing <PF18> (or <ALT-F8> if you are using a PC terminal). If you are denied access, you must contact the DB Administrator to get your privileges changed. If you are not denied access, call the Help Desk for consultation.

AFTER: (Exhibit 8.3.3.2b)

Exhibit 8.3.3.2b: Converting Prose to a Decision Tree

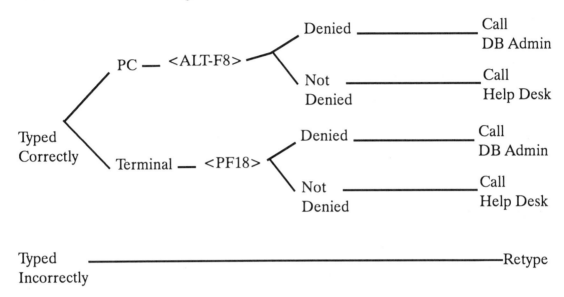

Notice how the passage below is converted to a decision table:

BEFORE:

Users may revise their passwords at any logon. After the first six months of employment, those with write privileges for the ATON database must revise their passwords at least once a week. Those with read-only privileges must revise their passwords at least once a month.

AFTER: (Exhibit 8.3.3.2c)

Exhibit 8.3.3.2c: Converting Prose to a Decision Table

More than 6 months employment	Y	Y	N	N
ATON Write privileges	Y	N	Y	N
Change Passwords at will			●	●
Change Passwords at least once/week	●			
Change Passwords at least once/month		●		

8.3.3.3 Handling Troublesome Procedures

If a procedure resists explanation by a competent writer, it is probably an error-prone procedure. Whether the procedure will get changed depends on the relationship between writer and developer.

Documentors who discover that they are trying to explain a very difficult (or nearly unexplainable) procedure have an interesting choice: either to proceed with the writing or to attempt to change the recalcitrant procedure. The better course is almost always the latter; the most productive changes include chunking the process into manageable pieces and improving the harmony between the physical and logical elements of the task.

What are the signs of an "unexplainable" procedure? Not only will it not fit into one module, it seems to want three or four levels of hierarchy in its explanation. Furthermore, when it comes time to partition the procedure into components or subprocesses, no logical or apparent pattern suggests itself. Or, worse, the competing patterns (such as segment-of-the-screen versus order-of-data-entry) are independent, unaligned.

Put simply, most procedures that resist a linear presentation—a simple sequence or hierarchy, with little skipping or looping—will resist being used. And valorous writers who take this complexity as a challenge to their writing skills are missing the point: Usability is everything. It is the better part of valor.

The form of communication between the developer of the procedure and the writer of the module (the documentor) depends on their working relationship. There are four common possibilities, and a fifth, uniquely effective arrangement.

Case I—The documentor is the developer. Although one would expect this to be the easiest case—the case most likely to result in a revision of the clumsy procedure—in fact it is one of the hardest. When the documentor is the developer, it usually means that a programmer has been conscripted into the documentor's role. Often, this person will be blind to the faults in the suspect procedure.

Case II—The documentor and developer are in the same group, under the same manager. This is one of the best arrangements; it presumes that both people are on the development team, committed to the most usable product possible. And unless the manager is one of those who value swift implementation ahead of quality, the bad procedure is likely to be changed. (Implement in haste; reprogram at leisure.)

Case III—The documentor and the developer work in two independent units of the same company or organization. (That is, the documentor's unit is not run by the developer's unit.) This is also a hopeful arrangement because it brings into play the inter-unit rivalries that frequently inspire innovations and explicit confrontations about quality. In the best case, the top management of the company becomes involved; presumably the conflict will be resolved in the way that best serves the organization.

Case IV—The documentor is writing up a procedure developed by *another company*. Here, again, little can be done to change the procedure in question. Indeed, I suspect that one reason so many firms sell software with so little how-to documentation is that they fear what would happen if the customer actually saw how awkward and difficult the program is.

Case V (the most productive arrangement, in most instances)—As in Case II, the documentor and developer are working together as a unit on the system, but the unit is a task force or ad hoc team invented to speed and improve the installation of the product. These teams usually have a greater sense of urgency about the project and are more willing to do the hard work needed to improve quality.

Whatever arrangement you choose, know that the documentor who is isolated from the technical developers almost always fails—and usually gets the blame for clumsy systems!

Exhibit 8.3.3.3: Five Paradigms for Documentors

Case	Arrangement	Effect
I	Documentor IS Developer	Deceptive; Too many blind spots
II	Documentor and Developer in same group	Productive; Common objectives
III	Two Units of the same company	Productive; Inter-unit rivalries
IV	Documentor writes-up 3rd party product	Frustrating; Hard to change the product
V	Task force, ad hoc team	Best; Focus on quality

8.3.4 Designing Reference Modules That Work

The type of documentation that benefits least from the modular format is reference *material: lists, inventories, and compendiums to be "looked up" as needed. The sole criterion for deciding whether to break a reference section into modules is whether this would make the material easier to find and use. If chunking the material does not aid the reference function, do not do it.*

Reference modules give reference—not teaching, motivation, or guidance. The reference function is to extend the memory of the user: to provide an accessible location for long lists of items that no one ever bothers to memorize, or a convenient access to items that were learned earlier but since forgotten.

As Exhibit 8.3.4 shows, the first task in designing a reference module or series of them is to assess the suitability of the "standard" presentation, that is, the typical method of presenting long lists and inventories.

Should the list be allowed to "wrap around," as the word processing literature puts it, or should it be modularized? For example, is there any advantage in recasting the most familiar reference material—a telephone directory—into two-page modules?

In many cases there is no advantage. I have seen "logical groupings" of reference lists that worked against the convenience of the reader. For example, one system with coded error messages divided the reference materials into "errors caused by the operator" and "errors caused by system malfunctions." Unfortunately, though, the operator could not recognize the class of the error from what appeared on the screen and often had to look in both places!

Well-designed reference modules do *not* try to teach. One of the earmarks of such a module is that it calls for a very short headline, a very short summary, and, often, no other text besides the summary. A typical reference module, when finished, will contain nearly two pages of exhibits (charts, tables, lists)

A manual full of reference materials—modular or otherwise—is probably not an effective user manual. Reference is what users need *after they know how to work the system or product.* Until then, reference material is often unfriendly or intimidating.

The most serious violation of this principle is the attempt to teach in a glossary. When a manual has been written for particular users, they should not have to consult the glossary (which is probably at the front or back) each time a new term is introduced. Glossaries are to help people remember what they have been taught in orientation or guidance modules. Sending readers to a glossary, or assuming that they will go there frequently, is a way of telling them that this manual was designed for someone else.

Reference modules alone cannot teach. Nor should they be embedded inside of teaching materials. It is inconvenient in the extreme for the experienced user to search an instructional section in pursuit of a frequently used table.

Rather, reference modules should always be easy to find. They should be at the beginning or end of the manual, even on the covers or the binder. They can be in the form of posters or pull-outs or pages that can be folded pocket-size for easy reference.

Operators often create their own reference materials and keep them in a tiny notebook or even taped to the underside of keyboards. If your users and operators are making their own reference documentation—and if you want it to be accurate and maintainable—you had better find out what they need and give it to them.

Exhibit 8.3.4: Designing a Reference Module

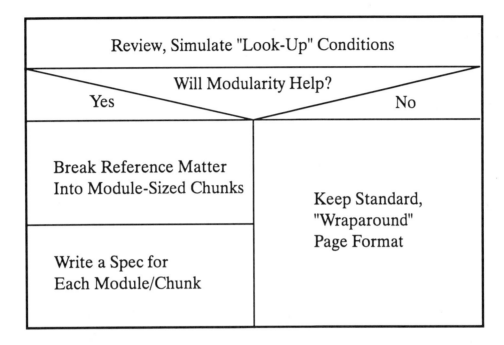

8.4 Mounting the Storyboard

The module specs are in a pile—an unworkable form for people who want to test and manipulate them. The next step, then, is to post them on a wall. In this form they can be reviewed and revised by the people who wrote them, the "authors" who will complete them, and the groups who will use the finished document.

The individual module specs are converted to a "gallery" by posting them in the intended sequence on the walls of a room. This process, converting the outline of a book to a visual display, is usually called "storyboarding," a term borrowed from the motion picture industry. (Interestingly, a technique suitable for planning movies is especially suitable for forcing documentors to think of their publications as sequences of information rather than hierarchical collections.)

In this form, the people who wrote the specs can really see them for the first time. They "walk through" the gallery, asking each other questions, challenging the emphasis, the scope, and the sequence of the several modules.

Then the "authors"—all the people who will contribute the missing details to the text and exhibits in the modules—are invited to review the storyboard and make further corrections or suggestions.

Once the planners and authors are satisfied, it is time for actual users to review the storyboard. The designers of the document should be present when the users or operators (or their representatives) review the storybord. The questions asked will reveal flaws in the design and may also correct misimpressions about what the intended readers actually know or do. A storyboard version of a user manual, if prepared early enough in the system development cycle, can actually point out ways to improve the design of the system!

The documentors should also *watch* the users and other readers as they review the plan. Often designers of the book can spot problems merely by observing the physical movements of the reviewers. Many of the design flaws of the book—loops and detours—will be evident as the users follow the logic of the manual, while there is still time to redesign.

For the full benefits of storyboarding to be realized, there should be *one* storyboard, posted in *one* place. In most organizations that is not a problem, but in some larger organizations, the various reviewers interested in the emerging document are at several scattered sites.

Though sympathetic to these problems, I still believe that there should be *one* storyboard, in *one* location. Innovations in networking and "groupware" notwithstanding, I recommend against having more than one review copy of any technical publication, and also against reviews carried on through the mails. The only thorough technical reviews I have seen were done with all parties present, with lots of questions and discussion, and with all the necessary people and data close at hand.

Eventually, when the designers are satisfied that all the valuable changes have been incorporated, they sign-off the design and invite an official (or official committee) of the organization to review the storyboard. If the design is approved, it is then frozen. That is, any proposed changes that will affect more than one module, must send you back to the storyboard.

Exhibit 8.4: The Gallery of Module Specs

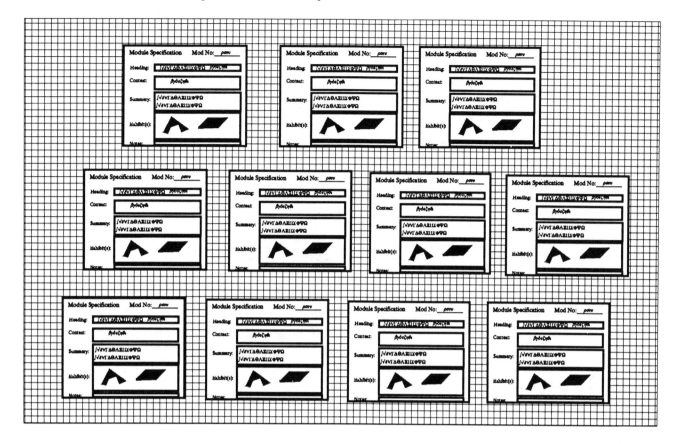

8.5 Modifying the Storyboard

As in structured design, and as in motion picture planning, one of the main benefits of the storyboard is the ease with which it can be changed. The irony is that the full spectrum of technical and communication flaws found in outlines can be addressed with relatively few "design moves."

A central purpose of the storyboard review is to control the number of loops and detours in the manual—GOTOs, in a figurative sense. Sometimes, we need nothing more than such an informal notion or constraint. In an informal review, the goal is to please the reviewers, rather than to meet formal standards.

In a formal review, though, there must be explicit criteria—especially when there is some dispute about the best design or sequence. As a central criterion, consider the following definition:

In a GOTO-less publication or manual, a reader who begins to read a module will finish reading it. If the reader needs or wants more information, he or she will move to the beginning of a new module...and finish that one. Moreover, in the most typical case, the second module will immediately follow the first.

In short, this constraint prevents documentors and writers from allowing or compelling readers to leave a module in the middle or to enter a module in its middle. It especially prohibits documentors from sending a reader from the middle of one module into the middle of another and back to the exit point in the first one.

Of course, the more diverse the audience for a particular manual or book, the harder it is to predict the ways in which its diverse readers will use it. Consequently, it is impossible in principle to develop a book that meets these criteria for all readers. And the greater the diversity of readers, the harder the task.

When possible, however, designers of the manual should recast and rearrange the module specs in such a way that the emerging document comes as close as possible to this standard. Every time the book forces a reader to exit or enter a module at the wrong place, you must try to change the design. (Once the first draft is written, it will be too late.)

And, surprisingly, even the most complicated changes can be handled with just a few "moves":

- **"DECOMPOSE"** (disaggregate)—Convert one module into two or more, in sequence or in hierarchy, with a new spec for each
- **CONSOLIDATE**—Collapse two or more modules into one, when they are part of the same theme or concept
- **INSERT**—Add one or more modules needed to bridge a gap
- **DELETE**—Change the sequence of two modules, from "logical" to "readable"
- **RELOCATE**—Move a module or group of modules from one place in the book to another

These moves account for most of the possible changes. (There are also changes within modules, which can be effected just by changing the contents of the module spec slightly, or by adding notes of emphasis.)

The storyboard technique was invented to ease the process of change and revision. A storyboard plan, unlike a text draft, can be revised radically a dozen times in a day. Even though modern document processors make it easier than ever to revise a "finished" draft, a full first draft will probably be only patched and plugged, never really redesigned to eliminate its flaws.

Exhibit 8.5: Storyboard Logic

BEFORE	PROCESS	AFTER
A	DECOMPOSE	A$_1$ → A$_2$ → A$_3$
A → B	CONSOLIDATE	AB
A → B	INSERT	A → X → B
A → B → C	DELETE	A → C
A → B ...→ X	RELOCATE	X → A → B

8.6 Won't There Be a Lot of Redundancy?

Ironically, one of the surest signs of success in writing a modular publication is that readers notice—or even complain about—the redundancy in the manual. Redundancy across *modules reduces the need to branch, loop, or detour. Redundancy* within *the modules compensates for noise and careless reading. Redundancy simplifies and reinforces.*

To sugarcoat the pill somewhat, I could have used some word other than *redundancy*: something like *repetition*, or *amplification*, or *reinforcement*, or *restatement*. Except for engineers, almost no one uses the term redundancy as a term of praise.

But redundancy is what it is: using more than is necessary, spending more than is necessary, writing equivalent information three or four or five times.

The redundancy in a usable user manual is of two kinds: **across modules** and **within modules**. Exhibit 8.6 demonstrates a simple kind of cross-module redundancy. There is a certain procedure that is at the beginning of several other procedures. In a nonredundant publication, the readers would be sent to the initial procedure again and again. (Before learning how to complete Task B, they would be told to read Task A; and the same for Tasks C, D, and E.)

But in a redundant publication, each later procedure would include an embedded explanation of the startup procedure, repeated identically each time. This practice is familiar, for example, in the manuals for calculators, which usually begin each procedure with a reminder to turn the calculator on and clear its registers.

This issue is complicated and controversial. What is especially interesting is that it suggests a breakdown in the analogy between modular computer programs and modular publications. In certain views, it is the essence of a structured program that it is *not* redundant, that whenever a

particular task or function occurs it is *called from the one place in the program where it resides.*

On closer examination, though, the analogy holds up. The real issue is whether the manual presented to the users has all the "calls" performed for them, or whether the reader is expected to search for the appropriate modules and run them at the appropriate times. And the factors affecting the decision are analogous as well. If the recurring material is rather large, it is undesirable to repeat it within each module. (In fact, it might take up all the available space.) If it is going to be invoked or referred to repeatedly, it would add too much complexity and difficulty to the reading process, in much the way that frequent calls add overhead to a computer program.

Redundancy, although it complicates maintenance and seems inefficient and wasteful, reduces the number of skips, jumps, branches, and loops in a publication. For readers with limited book skills, redundancy may be the difference between a usable and unusable book. And it follows, then, that developers of user manuals may feel safer putting less redundancy into those books intended for sophisticated users who handle complicated publications well. (Too much redundancy can irritate; excessively repeated directions sound preachy, especially when they review the basics.)

Redundancy across modules should use *identical* repetition. If a procedure or message or explanation is repeated, it should be repeated identically. Repeated paragraphs should be

indexed in the document database, so that they can be copied exactly and so that *when they are changed in one place they will be changed in every place they appear.*

Redundancy within the module compensates for noise (and inattentiveness): the headline is redundant with the summary, the exhibits illustrate the summary, and the text amplifies them all.

Exhibit 8.6: Redundant Modules vs. Called Modules

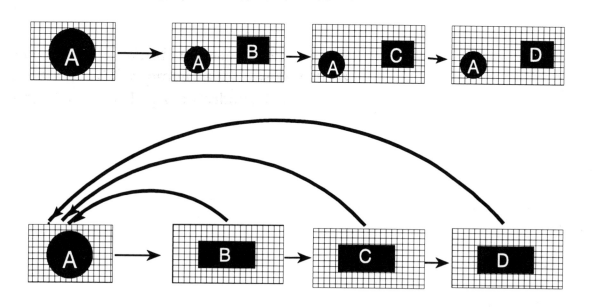

8.7 Handling Branches and Hierarchies

Is it logical to design a document as a series of independent modules, when so many tasks and processes have branches and paths? How does such a publication manage hierarchical procedures?

The suggestion that modular publications are an inadequate way to explain branching or hierarchical processes is ironic. Remember that *all* traditional manuals are a series of two-page spreads, except that the break of the pages is largely accidental. *All traditional user manuals, despite what some hypertext enthusiasts say, are hierarchical messages—networks of text nodes—that happen to be distributed as a series of two-page spreads.*

In short, the modular publication can present any logic that a traditional book can present. And probably better. Although the most usable manual will contain its branching procedures within *one* module, processes that cannot be so contained can simply branch to other modules. (It follows, then, that well-designed modules of documentation will *end at the branching point*.)

Sometimes, the writer wants the book to be read in a straight line (as in a proposal or tutorial); in those instances, the table of contents is presented without subordination, thereby discouraging users from reading it out of order. This device may even be judged a stratagem that prevents readers from ignoring the author's intended sequence. Consider this "straight-through" design:

Sequential (One-Tier) Outline:
    Copying the Distribution Disks
    Telling the System Your Configuration
    Choosing Options and Alternatives
    Setting Up A Mailing List
    Entering Data into the Mailing List
    Revising Data in the Mailing List
    Forming a New List with Parts of Other
      Lists
    Printing the Entire List
    Printing Selected Parts of the List
    Printing Envelopes
    Printing Labels
    Troubleshooting Chart

Alternatively, indentation and numbering schemes can be used to help readers find sections within the document.

Two-Tiered Hierarchical Outline:
    Four Steps to Get Started
      Copying the Distribution Disks
      Telling the System Your Configuration
      Choosing Options and Alternatives
      Setting Up A Mailing List
    Three Ways to Enter an Address
      Entering First Data into the Mailing List
      Revising Old Data in the Mailing List
      Forming a New List with Parts of Other
        Lists
    Printing the List
      Selecting the Addresses to Be Printed
      Printing Envelopes
      Printing Labels
      Troubleshooting Chart

Three-Tiered Hierarchical Outline

Note also that, as in this book, the headings at the top of a module reconstruct the "information stack" above the module. This device makes modular publications even more accessible than traditional ones; the hierarchy of the modular document is actually easier to see.

9. ASSEMBLY: GENERATING THE DRAFT

9.1 The Advantages of a Frozen, GOTO-less Design
9.2 Selecting and Managing "Authors"
9.3 Using Project Management to Assemble the First Draft

9.1 The Advantages of a Frozen, GOTO-less Design

The main beneficiaries of a GOTO-less publication are the readers. Additionally, though, the people who manage and write publications benefit as well. A GOTO-less design ensures the independence of the modules from one another, allowing them to be written in any sequence, by any arrangement of "authors"; and also allowing them to be reviewed and tested as they come in, without regard for sequence or for the links between the modules.

"Freezing a design," as explained earlier, does not mean that the design of the manual will never change. Rather, it means that the design is official and cannot be revised without an official routine. No one may make that small change which wrecks the GOTO-less design and, in the process, undermines the independence of the modules.

The GOTO-less manual is a collection of modules in which all the possible connections between modules—all the references and couplings—can be seen in the design. *Everything that writer A needs to know about writer B's module is already in the storyboard, in the module spec.*

Again, this functional independence among the modules can be lost in an instant if someone departs from, or adds to, the original design without also reworking the storyboard. (Changes that fit entirely within one module and do not affect any of the others are, of course, permitted. Generally, any version of the module that does not call for a new headline or summary paragraph is permitted at the discretion of the writer.)

An author—anyone assigned to write the body of one or more modules—can write the modules assigned to him or her in any sequence that is comfortable. A missing item usually cannot delay more than one module; the rest can be written independently.

And, as useful as the GOTO-less design is to the writing of the first draft, it is even more useful in the reviewing and editing of those modules. Put simply, once we know what is in the storyboard, we know enough to *review and edit any one of the drafted modules.* And if we do not anticipate any changes in the design, we can even assign page numbers and figure numbers to the modules, no matter the order in which they arrive. Instead of producing a long, tangled series of interwoven paragraphs, the writers produce a series of small self-contained publications, each of which has already been reviewed for its technical content, and each of which fits not only into the logic of the book but into the physical form of the book. So, the writers are *implementing, not creating.*

Developing documentation in this structured style reduces the interest of the first draft. Instead of being the most complicated, demanding, and fascinating part of documentation, writing the draft becomes the least interesting part. (Remember, most of the art and intellect has been shifted to the design phase.) Be warned, then, that even though books written by one person still benefit from the structured method, a professional writer will find it boring to carry out his or her own design and may be tempted to wander off onto artistic sidetracks.

The best plan is for two people, a technical expert and a documentor/editor, to design the publication and then to assign others the job of carrying out the plan.

Exhibit 9.1: Monolithic Documents vs. Modular

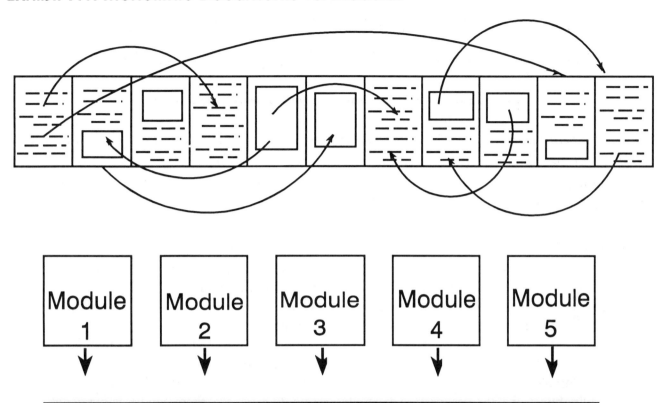

9.2 Selecting and Managing "Authors"

In the structured approach to user documentation, the first draft is merely the elaboration of the storyboard design. In the traditional approach to publication, however, the first draft is the first true attempt to organize and present the content. This difference changes the traditional conception of what "authors" do.

The similarity of programming to documenting is both inescapable and instructive. The programmer who works without structured analysis and design goes directly from a vague or intuitive notion to the program code. And most documentors—even professional technical writers—make the same mistake.

In the traditional, unstructured approach to manuals and publications, the first draft is one whole piece. At best, it is a few large pieces. Therefore, it is usually considered the large assignment of a single writer or, at best, the collaboration of very few writers. Because there is so little real specification in the conventional outline, logic demands that the many intricate connections among the parts of the publication be realized by a single person. Lacking external controls, the manual needs the internal control of a single author's mind to assure that all the parts hang together.

Unless the publication is divided into very large chunks (almost separate manuals), it is nearly impossible to do the work as a group. When the work is divided finely, however, the small parts rarely fit together. The problem is analogous to the problem of incompatible coding styles that plagued programming before the ego-less era of structured design.

In contrast, when the publication is fully specified in a set of module specifications, with each module small and independent, and each spec containing all important matters of technical substance, then writing the first draft is an entirely different task. Furthermore, when the

sequence of the modules is restricted by a GOTO-less logic, and when the design of the book is frozen, then the writing of the first draft is hardly like what is ordinarily thought of as "writing" at all.

The draft of the emerging manual is produced by having several authors supply the missing details in the text and exhibits—one module at a time. In principle, there can be as many authors as modules, each working independently. With this approach, the first draft of a very long manual could be completed by a team of authors within two or three hours of finishing the storyboard!

Even when there are only one or two authors, though, the benefits of the modular design are still impressive. The publication can be prepared in small installments, out of sequence, without worrying about the connections (the interfaces) across the modules.

Modular design also encourages full participation in the writing, even by those who are usually the most reluctant to write. In this scheme, the author is asked to provide correct details, within a prescribed space, for material that has already been designed (and reviewed and approved). Anyone who will not write under these circumstances—especially when told that grammar doesn't count—probably would not write under any conditions.

Using these unlikely authors not only allows for the rapid completion of the draft, but also improves the technical accuracy of the draft. If the writing is by the most knowledgeable person, the result, though awkward in style, is likely to be accurate. And it also liberates the professional writers to do what they do best: correct, clarify, and improve the writing of the drafts.

Exhibit 9.2: Managing Authors

Mod #	Apprv'd	Assigned Author	Draft Complete	Style Review	Tech Review	Final Approval
1	8/25/90	Gillis	✓	✓		
2	7/5/90	Gillis	✓	✓	✓	
3	7/5/90	Krebs	✓			
3.1	12/8/90	Osborne	✓	✓	✓	✓
3.2	12/8/90	Gilroy	✓	✓	✓	✓
3.2.1	12/8/90	Gilroy	✓	✓		
3.2.1.1	10/1/90	Krebs				
4	10/1/90	Gillis	✓	✓	✓	
5	8/25/90	Menninger				

9.3 Using Project Management to Assemble the First Draft

The benefits of modular documentation escalate rapidly as the size of the project grows or as the number of participants increases. Also, the management of writing is improved, first, by leveling the effort throughout the drafting stage, and, second, by allowing the documentor to use a full range of project management techniques.

Structured user documentation transforms the nature of assembling a first draft and, in the process, turns documentors into managers.

On any document big enough for two writers, someone must be in charge of the project. But in most organizations that produce documentation, there is very little real management. Working from conventional outlines, the documentors have no control over the time and cost of production—and limited control over the quality.

Exhibit 9.3a contrasts the traditional method with the structured method. In the traditional

Exhibit 9.3a: Comparative Distribution of Writing/Editing Effort

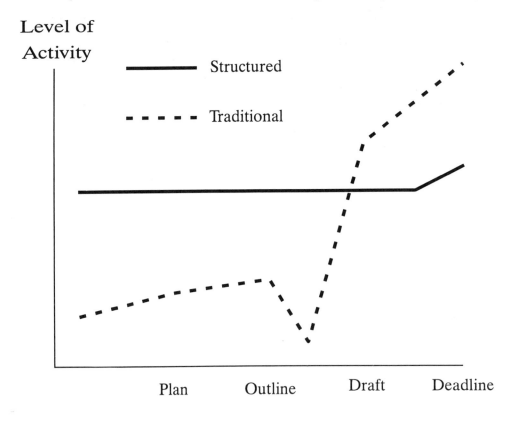

approach, there is a short period of outlining and planning, followed by a long trough of inactivity. During this interval, programmers and writers, who have usually been given writing assignments of undefined scope and size, stall and procrastinate until the deadline—or beyond. Meanwhile, the person in charge hopes ardently to get back some drafts, often without result.

Naturally, the person in charge ends up doing much of the writing alone, usually on a crash schedule, with little opportunity to edit and revise.

In contrast, a structured document takes longer to plan and design. But within only a few hours of finishing the outline, the first draft versions of the modules must come back to those in charge. There is a nearly level effort throughout the process, and time to edit, test, and revise.

Moreover, if a particular author fails to respond in the time allotted for writing a module, the documentor can investigate at once—perhaps

with the result that the module is assigned to another author, while there is still plenty of time.

This shifting of the production paths demonstrates another advantage of structured documentation: each module is a well-defined parcel of work and can be placed in a project network. (See Exhibit 9.3b.) Each module is a task of defined size, with a person in charge, an estimated duration, and, in some cases, a budget for artwork and production. The manager can estimate the costs beforehand and, by manipulating the assignment of modules, can predict and adjust the completion date. In the best case, all the modules are independent, so that the only constraint in the network is the result of having one author write more than one module.

The more thoroughly enforced the modular design, the greater the opportunities to employ project management tools, honor budgets, and shorten the "critical path" of the production.

Exhibit 9.3b: Allocating the Modules

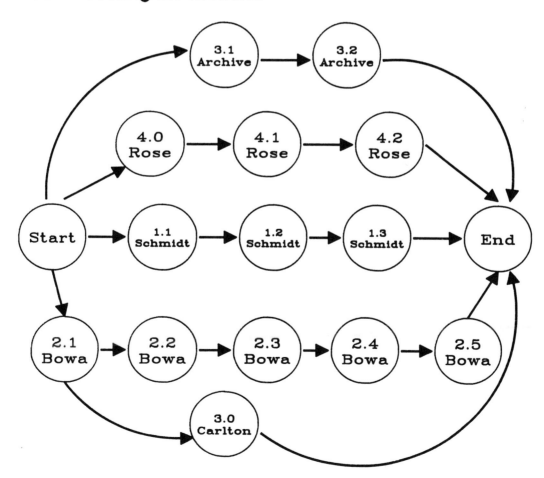

10. EDITING: REVISING FOR READABILITY AND CLARITY

10.1 Assessing the Draft: Main Issues

The purpose of assessing the draft is to correct the tactical errors. The goal is to rework as many as possible of the awkward and incorrect passages that cause readers to make mistakes or to reread the difficult or ambiguous sentences.

A monolithic first draft, written from a traditional outline or specification, is usually so difficult to read, and so filled with new technical material, that it is in urgent need of technical review. If time permits, there will be a style review and another quick technical review. A structured manual, however, can be reviewed module-by-module. Because the technical content of the module has already been reviewed, it is possible to clean up the language and presentation first, and then later do a light technical review to catch the matters of detail.

The traditional first draft of a document is mostly new—material that no responsible person has reviewed or tested before. Under these conditions, the logical thing to do is to begin with a meticulous review of the technical accuracy of the draft—a process made difficult by the unedited, first-draft prose. There is virtually no opportunity to make the language or artwork clearer, more readable, or more effective.

In contrast, each module of a structured manual is a self-contained micro-manual. The technical content and logical connections have been have been specified by a technical expert and reviewed by other experts. Because there are no big technical surprises in the module, and because the review is not delayed to the last moment, it is wise to begin with a language and art review. *It is easier to find technical errors in a clearly written (that is, edited) text than in a first draft.*

The purpose of this language review is to assure that the document is clear, free from ambiguity and misleading information, and readable. It should be no harder to read than it must be, and certainly not too hard for the intended reader. In general, the more resources spent on editing a draft, the clearer and more readable it becomes.

The editorial improvements are usually in five categories:

- **Mechanics**—correction of errors of usage, grammar, spelling, and punctuation
- **Appropriateness of language**—replacement of words and phrases that are unfamiliar to the reader, unnecessarily difficult, wrong in tone
- **Clarity**—replacement of words, phrases, constructions, or graphics that have several possible meanings, or that mislead the reader
- **Accessibility**—elimination of awkward, show-off, wordy constructions and difficult artwork; elimination of the first-draft commonplaces, such as backward sentences
- **Urgency**—revision to make the writing more interesting and engaging, through careful diction, close editing, variations in sentence length and style, and prose skill

If someone in the organization complains about the time spent on editing, explain that unclear sentences mask technical errors and invite trouble. For example, the sentence

Managers are required to sign off on Form A51 to approve continuation of a project.

is a horrible mess. And the worst thing about it is that it seems to oblige managers to sign a particu-

lar form—which is not intended. Applying
principles explained later, you may convert the
sentence to

If managers want to continue the project, they
must sign Form A51.

Exhibit 10.1: Alternative Ways to Edit a Draft

IF THE DRAFT IS MONOLITHIC...	IF THE DRAFT IS MODULAR...
Then, for the Whole First Draft	Then, for Each Module

Initial Technical Review

- first reading of complicated text
- discovery of major errors
- attempt to separate technical errors from ambiguous language

Initial Language Review

- revision of raw draft language
- clarification of ambiguous content
- identification of gaps and inconsistencies

Quick Language Review

- hasty review of mechanics
- minor editorial cleanup
- writing specs for production

First Technical Review

- verification of technical content NOT already tested in storyboard
- incorporation of late technical changes

Final Technical Review

- rushed, perfunctory double-check
- clumsy updating (errata pages)

Final Language Review

- refining language/art for usability
- careful production and proofing

10.2 Editing for Word and Phrase Bugs

The easiest improvement to make is removing or revising certain common word and phrase bugs. Notable among these are showing off with long or fashionable words, using too many words, using too few words, and putting words and phrases in the wrong places.

Certain recurring errors of style have the double effect of, first, masking the technical errors and omissions in the text, and, second, increasing the chances that the reader will reread or misread your explanations.

Showing off consists in using long words where short, familiar words would have been just as effective. For instance, *utilize* for *use, facilitate* for *help, initiation* for *start,* even *depress* for *press*.

Do not misunderstand. There is no reason to write in one-syllable words, and there is no advantage in replacing a technically correct word with a shorter, incorrect word. But there is no gain in using *indicate* for *show,* or *disseminate* for *spread,* or *effectuate* for *cause*.

Another common form of showing off is the use of **vogue words** (buzzwords), like *capability* for *ability,* or *prioritize* for *rank.* The word *environment* appears so often that I have seen it used with opposite meanings in the same publication. Also beware of *transparent* which means "invisible" to computer people and "obvious" to speakers of business English.

Using too many words is a technique youngsters learn as a way of stretching a 300-word idea into a 1000-word composition. A few examples will demonstrate:

- *should it prove to be the case that = if*
- *by means of the utilization of = with*
- *at that earlier point in time = then*
- *conduct an inspection of = inspect*

- *perform the calculation of the projections = project*

Wordiness is nearly inevitable in first drafts. The most frequent offenders are the "smothered verb" (*make a distinction* for *distinguish, accomplish linkage between* for *link*); the phrase used where a single word would do (*in order to* for *to, with regard to the subject of* for *about*); and, occasionally, clauses for phrases:

BEFORE: *After we had approved the test plan, we began the...*

AFTER: *Having approved the test plan, we began the...*

Using too few words is found often in the writing of engineers and computer programmers. Driven by a desire to be concise, some writers produce phrases and sentences that are compressed to the point of incomprehensibility. What does it mean, for example, to say that a certain system has an "English-like report generating capability"? What are "contiguous sector reference designators"?

No one but the author is sure what is meant by "operational planning materials format design criteria" or "management responsibility assignment history file." Most people cannot understand strings of nouns, or what the programmer would call "noun strings." And adding a few modifiers does not help.

People who avoid prepositions and who cram words together in this cryptic way also tend to eliminate other "useless" words like *the* and *a*. If a *that* is optional, they remove it. Indeed, they tend to leave out all the optional commas as well. The trouble with these zealous choppers and cutters is that they destroy the flow of the language and produce sentences that, though they may be shorter, probably take longer to read.

Misplacing words and phrases can also throw your readers off the track. In English, modifiers should be next to the words they modify—usually before. But in most first drafts, many of the modifiers tend to be misplaced, notably *only, nearly, almost, already, even*, and *just*. In the instruction

> *Only* enter the hourly rate for exempt employees.

there are various ways to interpret the only. (Only the *hourly* rate; the hourly rate and *nothing else*; or the hourly rate *only for exempt employees*.) Be careful of these modifiers. Do not write

> The system *nearly prints* everyone's checks.

when you mean

> The system prints *nearly everyone's* checks.

Similarly, descriptive phrases should be near the word or phrase they describe. What does this instruction mean?

> Report every unauthorized access in keeping with company policy?

Does this mean that if the unauthorized access is *not* in keeping with company policy that you should not report it?

10.3 Editing for Sentence Bugs

Although there are scores of things that can go wrong with a sentence, the five flaws most likely to stop or distract a reader are backwards construction; meaningless predicates; tangled passives, dangling introductory phrases; and marathons.

The secret of the readable sentence is that the "payload" of the sentence—the material that the author would have underscored (if something had to be underscored)—is *at the end*. If a sentence is long, it is read and processed in stages; the last read part is the best remembered part.

Any writer can learn the drill: review the first draft of the sentence; see if the material to be emphasized is at the end; if not, rework the sentence to move it to the end—unless there is some technical reason for not doing so. So, we convert

Reduced cost is the main advantage of this new procedure.

to

The main advantage of this new procedure is reduced cost.

Similarly, when we edit instructions, we put the key material last. And if it is a conditional instruction (*if-then*), we make sure the *then* clause is last.

No: DFIL is typed.
Yes: Type DFIL.

No: Type DFIL to see what file names have been assigned.
Yes: To see what file names have been assigned, type DFIL.

If the payload of a sentence is at the end, then it follows that the "action" in the sentence is usually in the predicate, not the subject. Yet, not only do many writers put their main material at the beginning, they sometimes say everything interesting before they even get to the verb, leaving a **meaningless predicate**.

Consider: "The possibility of underpricing by the Japanese exists." The entire predicate of the sentence is the word *exists*. But to revise the sentence we have to know what the writer wants us to understand. Is it (1) "The Japanese may underprice us." or (2) "We may be underpriced by the Japanese."? Both sentences are grammatically correct. Sentence 1 emphasizes *us*; sentence 2 emphasizes *Japanese*.

English is filled with devices that allow the editor to move phrases from front to back. Among the most useful is the passive form of the verb. Converting (1) "ZAKO Industries acquired an XTRON." to (2) "An XTRON was acquired by ZAKO Industries." changes the verb from active to passive and changes the emphasized word!

Most editors and teachers of writing warn against the **passive form**—with just cause. Tangled passives can ruin an otherwise understandable passage. Consider these pairs:

Passive: Insufficient flexibility is exhibited by the system.
Active: The system is too inflexible.

Passive: Cheap collating and binding are accomplished by this device.
Active: This device collates and binds cheaply.

Passive constructions are typically wordy and difficult. But they can, when used carefully, help you to propel the payload of a sentence to the most effective position: the end.

Still another way to push the main stuff to the end is to use introductory phrases. (Nearly all conditional instructions have introductory phrases.) The danger in these is the **dangling introductory phrase**, a string of words disconnected from the body of the sentence.

Again, the drill is simple. The introductory phrase must be tied to the grammatical subject, which should appear right after the comma.

No: With your simple payroll requirements, PAAY is the system for you.

Yes: With your simple payroll requirements, you should use the PAAY system.

No: To locate definitions quickly, glossaries are posted at each work station.

Yes: To locate definitions quickly, operators can use the glossaries posted at each work station.

No: When coldstarting the system, the operating system tape is loaded.

Yes: When coldstarting the system, (you) load the operating system tape.

There are also some strange danglers at the ends of sentences. Beware of such absurdities as: "Do not service the printers while smoking."

Finally, someone must be sure that the sentences simply do not run on too long. The problem is with long sentences in general, especially with several in a row. No one can handle the **marathon sentence** below:

In addition to solid, dashed, phantom, centerline, and invisible line fonts, numerous linestring fonts are available that provide generation about a centerline with variable spacing (width), layer of insertion options, and left, right, and center justifications.

10.3.1 Nine Ways to Write an Unclear Instruction

Any word, phrase, or sentence bug can hurt clarity and usability. And the consequences of unclear instructions can be expensive.

1. **Long, vogue words.**

   BEFORE: In the Information Center environment, the manager should utilize a prioritization ranking to facilitate equitable scheduling.

   AFTER: In the Information Center, the manager ranks each job to yield a fair schedule.

   BEFORE: If your configuration has sufficient RAM capacity, you may utilize the system's windowing capability.

   AFTER: If your computer has enough memory, you can use the *window* feature.

2. **Too many words.**

   BEFORE: In the event that you have a lack of knowledge regarding which files you have permission to write in, make use of the PRIFIL command.

   AFTER: If you do not know which files you may write in, type PRIFIL.

   BEFORE: Should it prove to be the case that you have some reservations regarding the forecasts, you have the option of using alternate discount rates.

   AFTER: If you doubt the forecasts, try other discount rates.

3. **Too few words.**

   BEFORE: Column heading revision permission may be obtained by HCOL entry.

   AFTER: To get permission to change the headings of the columns, enter HCOL.

   BEFORE: Early manual design yields procedural usability benefits.

   AFTER: Writing manuals early makes the procedures easier to use.

4. **Misplaced words/phrases.**

   BEFORE: Only write corrections, not changes, on the worksheet.

   AFTER: On the worksheet, write only corrections, not changes.

   BEFORE: The slide-maker only can be used by systems with 512K memory and hard disks.

   AFTER: The slide-maker can be used only by systems with 512K memory and hard disks.

5. **Backwards construction.**

   BEFORE: Press the <Clear Rest> key if you want to erase everything after the cursor.

   AFTER: If you want to erase everything after the cursor, press the <Clear Rest> key.

   BEFORE: Type PINSTALL to change the printing options.

   AFTER: To change the printing options, type PINSTALL.

6. **Meaningless predicate.**

BEFORE: The efficiency of spot-checking the data sheets before commencing entry is worthy of mention.

AFTER: It is efficient to spot-check the data sheets before you enter the data.

BEFORE: The urgent need to save data at least every ten minutes is called to your attention.

AFTER: You must save the data at least every ten minutes.

7. **Tangled passive.**

BEFORE: Care must be exercised in sending sensitive data.

AFTER: Send sensitive data carefully.

BEFORE: File linkage can be accomplished by key specification.

AFTER: To link the files, specify the keys.

8. **Danglers.**

BEFORE: When reconciling the account, the encumbrance file must be frozen.

AFTER: When reconciling the account, (you must) freeze the encumbrance file.

BEFORE: To call the Calculator, <alt> and <c> must be pressed.

AFTER: To call the Calculator, (you) press <alt> and <c>.

9. **The unnecessary third person.**

BEFORE: The operator then enters his or her security status.

AFTER: Enter your security status.

BEFORE: The clerk should then type the number of the desired file.

AFTER: Type the number of the file you want.

10.3.2 Increasing the Power of Instructions

A common problem in writing instructions for users and operators is the overreliance on such words as responsibility *or* requirement *in place of the far clearer auxiliary verbs:* should, must, *and* shall.

User documentation is—or should be—filled with direct instructions, directives. Yet, certain long-winded and evasive habits of style undermine many such sentences. The following pairs of words deserve suspicion:

- requirement/required
- responsibility/responsible
- necessity/ necessary
- obligation/obligated
- mandatory/mandated

These words are "suspect"—not wrong or substandard. When they appear, there are likely to be two serious problems in the sentence. First, it is almost certainly wordy and unnecessarily hard to read. Second, the author's intention is ambiguous.

These suspect words (and others like them) are usually stuffy substitutes for the more powerful auxiliary verbs *should, ought to, must, has to,* or *shall*.

As Exhibit 10.3.2 shows, most directives have one of three levels of intensity. A **recommendation** is an urging; the writer wants readers to follow the instruction but does not insist. A **procedure** is more compelling; the writer wants readers to understand that not following the instruction constitutes an error. A **policy** (or contractual provision) is the most compelling; failure to follow means that there will be sanctions, penalties, or withheld payments.

Exhibit 10.3.2: Selecting Auxiliary Verbs for Procedures

Intention	*Auxiliary*
Recommendation	Should, Ought To
Procedure	Must, Has to
Policy, contract	Shall

Consider the following:

It is a *requirement* that operators receive 40 hours of instruction before they enter any real data.

Not only is the sentence garrulous and unreadable; it is, more important, unclear. What happens if operators do *not* receive 40 hours of instruction? Will the infraction be winked at (a recommendation ignored)? Will the training director be criticized for failure to follow the SOP? Will a payment be withheld (contract violation)?

The ambiguity is resolved by choosing one of the following:

Operators *should* receive 40 hours of instruction . . .

Operators *must* receive 40 hours of instruction . . .

Operators *shall* receive 40 hours of instruction . . .

Similarly, the sentence

It is the responsibility of the arriving operator to read the trouble report from the latest shift.

becomes

The arriving operator *ought* to read the trouble report from the latest shift. *or*

The arriving operator *has to* read the trouble report from the latest shift. *or*

The arriving operator *shall* read the trouble report from the latest shift.

Why do so many writers resist these clearer, simpler alternatives? In many cases the impulse to show off is coupled with the desire to be evasive; that is, not only do they want to use impressive bureaucratic terms (like *mandated*),

but they also, ironically, do *not* want to assert their claim with power or authority. There are whole organizations reluctant to tell people, unmistakably, what to do, especially when the readers are professionals. (One Canadian official told me that putting unambiguous procedural language in a policy manual would reduce the ministers to obedient clerks.)

Another complication is the near absence of the term *shall* in the writing of North Americans. Outside of the legal profession, few writers know the correct occasion for the word. (They just know that it sounds more ceremonial.)

In the second and third person (you shall, they shall), *shall* has the force of law. The ponderous

Users are obliged by law to read the copyright disclaimer.

becomes

Users *shall* read the copyright disclaimer.

Thus, depending on what is meant, the expression

Analysts are responsible for validating the spreadsheet formulas.

becomes

Analysts *should* validate the spreadsheet formulas. *or*

Analysts *must* validate the spreadsheet formulas. *or*

Analysts *shall* validate the spreadsheet formulas.

Note: When a document contains *only* procedures, you may dispense with the auxiliary verbs. The last example becomes "Analysts validate the spreadsheet formulas."

10.4 Making Text Easier to Read

The term readability *refers to the difficulty of a particular text. The word* difficulty *here refers to the sheer effort needed to read a passage. The most popular of many indexes of readability are Robert Gunning's Fog Index, a simple technique for approximating the "grade level of difficulty" of a passage, and the Flesch-Kincaid Readability Index.*

All readability scales are imprecise, and many scholars question their validity; probably any one of them can be faked. Everyone has seen ingeniously composed passages that scored "easy" on the readability scales but were, obviously, nearly impossible to read. The purpose of these metrics is to extract some "objective" assessment of how hard a passage is for the reader to process. The most popular scales usually contrive to have the score equal the "grade level" of difficulty, that is, the number of years of schooling needed to read the passage with ease.

The best known, Robert Gunning's Fog Index, adds the average number of words in a sentence to the percentage of "hard" words and multiplies by a constant (.4) to yield the Fog Index. (In Gunning's scheme, a "hard word" is any word with three or more syllables, except for proper names, compounds of simple words, and three-syllable words in which the third syllable is *ed* or *es*.)

Another popular measure, the Flesch-Kincaid Readability Index, is used by, among others, the U.S. military in testing the reading difficulty of its manuals. This favored military scale is a revision of the Rudolph Flesch READ scale, calibrated so that it, too, reports grade level. (*Note*: Most style-checking software programs compute one or both of these readability indexes.)

To test the readability scales, consider this passage, published by one of the world's largest manufacturers of hardware and software:

Today's advancements in educational management combined with the rapid growth in student enrollment in schools has emphasized the need for data processors to be used in establishing and maintaining a student records data base, required for providing attendance and academic mark reporting data to satisfy several disciplines. The purpose of this program product is to provide a systematic procedure for recording, retrieving, manipulating, and reporting significant student data, such as attendance and academic mark information. One of the objectives of this program is to provide effective data on individual students as well as aggregate, statistical reports needed for sound analytical decisions by educators and administrators.

This passage has 105 words, 3 sentences, and 34 "hard" words, according to the Gunning criteria. Its Fog Index is .4(35 + 32) = .4(67) = 26.8. The Flesch-Kincaid Index rates it 21.6.

There is no person on Earth who can read this passage without difficulty! And this is especially unfortunate when you realize that the passage says very little indeed, and could easily have been revised to the 10 or 11 level.

Of course, merely lowering the readability score of a passage does not solve all its problems. A text with a score of 6 or 7 can still be unintelligible. Whatever quarrel one might have with these particular indexes—or even with the entire concept of simple readability measurement—there is no denying that excessive difficulty assures that most readers will be unable to make sense of their manuals. Even if the documents are clear, correct, and well designed to eliminate GOTOs, they may still prove unreadable.

Even when the audience is presumed to be sophisticated and well educated, the simpler a technical publication, the more people there are who can read it.

Exhibit 10.4: Two Readability Formulas

The Fog Index (Gunning)

Grade Level of Difficulty =

.4[average words/sentence + percentage of hard* words]

* Hard Words =

all words with 3 or more syllables, except

- proper names
- compounds of small words
- 3-syllable words in which the third syllable is *ed* or *es*, which would otherwise have had only 2 syllables

---

The Flesch-Kincaid Index

Grade Level of Difficulty =

[.39(average words/sentence) +
11.8(average syllables/word)] - 15.59

10.5 Demonstration: Procedures Before and After

"Before" and "after" versions of two actual passages from user documents illustrate the effects of editing the draft. After several editorial improvements, each passage shows a dramatic reduction in reading difficulty, as measured by the Fog Index.

The passage below comes from a real manual, the project development guidelines for a large financial institution.

BEFORE:

Following identification of needs and appropriate preliminary approval for all major system development project proposals, the Information Systems Department will prepare an analysis and recommendation for action. The more routine requests will be approved by concurrence of the Information Systems Department and of the financial area management without further review. Those requiring a change in policy, exceeding the approved budgets or crossing organizational lines, will require review and approval by the Steering Committee as well.

The Information Systems Department will evaluate the capability of the user or regional technical staff to implement a proposed system. Based on this evaluation, the responsibilities and authorities of the Information Systems Department, regional technical staff, and the user will be outlined in a system development proposal submitted to the Steering Committee.

Words: 127   Sentences: 5   "Hard" words: 37
Fog Index: 21.6

With a bit of editing, an exceedingly difficult (though typical) bit of administrative procedure becomes easy enough for any business professional to follow.

AFTER:

First, needs are identified and major development proposals get preliminary approval. Then, the Information Systems Department analyzes each request and recommends an action.

For small, routine requests the Information Systems Department and the manager of the functional area may approve the project without further review. (A project is "routine" if it does not call for a change in policy, exceed current budgets, or cross organization lines.)

For major requests, though, the Steering Committee must also approve. To advise them, the Information Systems Department submits its own evaluation, which proposes schedules and tasks for all its participants.

Words: 97   Sentences: 6   "Hard" words: 12
Fog Index: 11.4

In the next case, the "before" comes from the FORTRAN programmer's guide published by a service company.

BEFORE:

It is critical that variables used as subscripts in FORTRAN programs always be consistent with information declared in the DIMENSION statements. Unless checking is specifically requested, subscript ranges are not checked for validity when programs are run. This checking is omitted in order to maximize running-time efficiency. However, if invalid values are used for subscript variables, such as a value less than one or greater than the maximum subscript as specified in the DIMENSION statement, errors can occur. Often such errors either go undetected or cause apparently unrelated failures and diagnostics.

When invoking the FORTRAN compiler, the user can inform the compiler that subscripts are checked for range validity by supplying the SUBCHK option.

Words: 115   Sentences: 6   "Hard" words: 21
Fog Index: 14.8

In the "before" form, this passage is understandable to a good reader after two or three attempts. The programmers who get it on one reading are those who already know what it means. The "after" version, however, without "talking down" to anyone, brings the material into the range of most of the English-speaking adults who might choose to read this passage.

AFTER:

Variables used as subscripts in FORTRAN programs must stay within the range of those in the DIMENSION statements. (That is, the value of the variable must not be less than 1 or greater than the highest subscript in the DIMENSION statement.) If they are out of range, invalid, the mistake is often overlooked. Worse, these errors often cause "unrelated" failures or odd diagnostic messages.

To save running time, this system does not check the range of the variables unless told to. To be safe, when you invoke the FORTRAN compiler, tell it to validate the values with the SUBCHK option.

Words: 100   Sentences: 6   "Hard" words: 8
Fog Index: 9.8

10.6 Other Ways to Make a Document More Accessible

To make a manual more accessible, documentors should eliminate as many distractions as possible, present the material in a package that communicates confidence, and lay out the pages effectively.

Each document should be freed from distractions, especially mechanical and production errors: mistakes in spelling, punctuation, or grammar; inconsistent conventions and terminology; acronyms and abbreviations that are not spelled out often enough; awkward layouts; text lines more than 5 inches wide, poor photography or color separation. Each occurrence of these bugs, though not likely to do much harm in itself, can distract and confuse just enough to undermine an instruction or break a reader's concentration. Moreover, a recurring pattern of such bugs can imply an attitude of carelessness or sloppiness. And that is simply the worst possible message.

We should do everything possible to communicate confidence to the reader. Careful editing for small bugs will help. So will high-quality printing, copying, and binding. Expensive paper may be the hardest aspect of documentation to justify, but it does, unquestionably, create a better response in users and customers than cheap paper.

If documents are printed on two sides, be sure that the paper is heavy enough so that the characters and graphics do not "bleed through" the back. If your copy machine is a "bargain," be sure your pages do not look like a "bargain."

Be warned that anything that looks cheap or chintzy may undermine the effectiveness of a document. Usually, it is just a matter of taking away the reader's respect: The user does not take seriously what the documentor did not take seriously. Often, though, the cheapness produces

material that is nearly inaccessible. For example, the practice of cramming as many words as possible onto a page—the refusal to use large, good fonts, highlighting, or any other form of more sophisticated desktop publishing—produces manuals that are torture.

Further, documentors must be wary of any manager whose principal objective seems to be to saving space. *There is no communication benefit in conserving paper.* Wide margins and big print are better for readers—all readers. An uncluttered page is a page less likely to produce fatigue, and, therefore, less likely to encourage errors. Thick paper, good binders, tabs between the sections, better typography, color—none is essential, but all can help a system realize its full usability.

Documentors must also be wary of the brand of editor whose objective seems to be to save paper by the reckless elimination of words and the incessant use of abbreviations and other compressed forms. There is a profound distinction between clear, concise writing, on the one hand, and compressed, impenetrable writing on the other. (An editor who would cut *on the one hand...on the other* from the last sentence does not understand this point.)

Ultimately, firms and organizations that produce lots of publications must acquire competent, professional editors. Programmers can be taught to write a little better; "style-checkers" can catch some mechanical errors and compute a Fog Index. But there is still a need for someone who knows that good communication demands patience and rewriting, someone who knows the difference between conciseness and denseness, between compactness and clutter. And, perhaps, someone who knows that the word "prioritization" is just plain ugly.

Exhibit 10.6: Saving Paper vs. Readability

CONSERVES PAPER	HELPS THE READER
Narrow Margins	Wide Margins
Small Type, Dense Layout	Larger, Varied Type Sizes
Few Illustrations and Exhibits	Frequent, Large Charts, Art…
Run-On, Wraparound Printing	New Page for Each New Section
Austerity, Slim Explanation	Redundancy, Accessibility Tools
"Typewriter" Headings	Typography/ Desktop Publishing
Compressed Graphics	Large, Full-Sized Graphics

10.7 Using Style-Checking Software

A relatively new tool for writers is style-checking software, programs that scan drafts for lapses of grammar and, more important, for common excesses and affectations of style. Though far from perfect, these programs are a great aid in proofreading.

"Style-checkers" are an ambitious extension of the "spell-checker." Instead of looking just for unfamiliar strings of characters (what spelling checkers really do), they also look for grammatically incorrect sequences ("He don't"), incomplete patterns (like unclosed parentheses), impossible punctuation (sentences without initial capitals), and similar problems that can be expressed as algorithms.

Where style-checkers are most interesting and controversial is in their application of "rules" of good writing, an area in which most amateur writers are reluctant to take instruction. Popular style-checkers will flag jargon and neologisms (like *prioritize* or *impact*), warn writers about commonly misused terms (*presently* or *effect*), scold them for sexism, hector them about long sentences, chide their colloquialisms, and suggest succinct replacements for verbose constructions. The most popular products will tag nearly every passive form of the verb—a boon for most technically oriented amateur writers.

Style-checkers also perform statistical appraisals of writing style. They calculate Fog or Flesch-Kincaid readability formulas, as well as numerous other indexes and distributions. (One program complains if nearly all the sentences start with the same part of speech.)

Style-checkers also make mistakes. They misidentify the beginnings and ends of sentences with some frequency. They often miss glaring errors of grammar, and, even more frustrating, they often call something wrong that is just fine. Most of these lapses are programming problems, errors in parsing sentences. But some are substantive errors: superstitions and misconceptions reminiscent of the false "rules" imposed by grade-school teachers—for instance, nervousness about starting sentences with conjunctions or ending them with prepositions. One popular program believes there is a rule against putting a single word in quotation marks; another thinks that paragraphs must have at least two sentences. Moreover, most style-checkers ignore or misinterpret elliptical material, like headings and captions.

Most of the professional writers I know are cynical toward these inexpensive software editors. Indeed, most good, careful writers learn little from using them and resent the shallow advice they receive.

But that misses the point. Style-checkers are not for good, careful writers; they are for careless and mediocre writers. Which is to say they are for most of the people who create corporate and government publications and screen messages.

Style-checkers provide low-level editing for people who otherwise would receive none. And they provide moderately good writing instruction for people who have had almost none.

Exhibit 10.7 shows the comments inserted into a sample document by the program RightWriter.

A second sweep of the document with the program Grammatik additionally tagged the need to replace "one of the" with a shorter form, to substitute *that* for *which* in "which are listed," and to replace "determine" with a simpler form.

Neither program catches the hideous error of agreement in the last sentence ("operator . . . they . . . their"), or the lack of helpful punctuation in that generally painful sentence. Moreover, neither program suggests that this procedure would be more readable in a three- or four-step itemized procedure, rather than in paragraph form.

In sum, even if all the advice from both programs were followed, there would still be serious problems. But even so, the reader would be much better off than with the unaltered original.

Exhibit 10.7: An Assessment by RightWriter

Sample

Prior to inserting a graphic into the document, it is required that the file format be determined. If the format is determined to be one of the bit-mapped types which are listed in Table A, then "Import" should be selected from the File Menu. If it is found to be one of the types in Table B, "Open" should be selected. The operator who can not determine the file type should contact their System Administrator who will either apprise them of the file type or alternatively make a recommendation that some graphics conversion program must be utilized.

RightWriter **Analysis**

Prior to inserting a graphic into the document, it is required that the file format
        < < * **U12. WORDY. REPLACE Prior to BY before** * > > ^
          < < * **S1. PASSIVE VOICE: is required** * > > ^
be determined. If the format is determined to be one of the bit-mapped types
    ^ < < * **S1. PASSIVE VOICE: be determined** * > >
  < < * **S1. PASSIVE VOICE: is determined** * > > ^
which are listed in Table A, then "Import" should be selected from the File
     ^ < < * **S1. PASSIVE VOICE: are listed** * > >
       < < * **S1. PASSIVE VOICE: be selected** * > > ^
Menu. If it is found to be one of the types in Table B, "Open" should be
   ^ < < * **G3. SPLIT INTO 2 SENTENCES?** * > >
   ^ < < * **S3. LONG SENTENCE: 28 WORDS** * > >
      ^ < < * **S1. PASSIVE VOICE: is found** * > >
selected. The operator who can not determine the file type should contact their
   ^ < < * **S1. PASSIVE VOICE: be selected** * > >
System Administrator who will either apprise them of the file type or
  < < * **S13. REPLACE apprise them BY SIMPLER let them know?** * > >
alternatively make a recommendation that some graphics conversion program
< < * **S15. IS THIS AMBIGUOUS? some graphics conversion program** * > > ^
must be utilized.
   ^ < < * **S1. PASSIVE VOICE: be utilized** * > >
  < < * **S13. REPLACE utilized BY FORM OF SIMPLER use?** * > >
    ^ < < * **G3. SPLIT INTO 2 SENTENCES?** * > >
    ^ < < * **S3. LONG SENTENCE: 36 WORDS** * > >

READABILITY INDEX (Flesch-Kincaid): 11.91

  Readers need a 12th grade level of education.

SENTENCE STRUCTURE RECOMMENDATIONS:
   1. **Most sentences contain multiple clauses.**
     Try to use more simple sentences.
   3. **Most sentences start with nouns.**
     Try varying the sentence starts.

11. TESTING: DEVELOPING A FORMAL USABILITY TEST

11.1 Elements in a Well-Made Usability Test

A usability test is any systematic, formal project whose aim is to gather reliable, generalizable data about the uses and usefulness of a product or publication. The methods currently in use range from formal laboratory experiments to anthropology-like field studies.

Nearly all professional technical writers agree: You cannot be sure that any procedural document is clear until it has been tested with appropriate readers.

Originally, the usability testing movement was meant to be an alternative to the informal judgments of writers and reviewers. Not surprisingly, then, its earliest advocates proposed an especially rigorous laboratory model. Today, though, usability testing employs all the tools of social and educational research, including contextual and longitudinal studies.

The variety of method is impressive. In the traditional approach, the investigators are unobtrusive; in context studies, they interact with the observed. In lab tests, the developers are excluded; in "wizard of oz" tests, the developers are manipulating the material seen on the test screen.

Usability research, apparently, is in the domain nowadays called *evaluative research*: methods somewhere on the spectrum between formal science and responsible journalism. Even under what the scientist would consider sloppy conditions, usability researchers gain insights that

Exhibit 11.1: Sample Test Objectives

Antecedent:	Given a menu of available printer drivers
Task:	The operator will be able to install a supported printer
Precision:	Within 2 minutes from the opening menu
Reliability:	80% of the time
Antecedent:	Given a list of incorrect addresses
Task:	The operator will be able to update the mailing list correctly
Precision:	At a rate of 40 addresses/hour
Reliability:	80% of the time
Antecedent:	Given the data for the table of organization
Task:	The user will be able to generate a camera-ready organization chart
Precision:	In under 30 minutes
Reliability:	90% of the time

could not have been obtained in the era when evaluation of manuals consisted in reading any feedback cards that may have been received.

Clearly, if the objectives are thoughtful, and the methods are sensible, a usability test adds information that is helpful to the developers. It is also clear, though, that done carelessly, a usability test can be manipulated to give apparent approval to mediocre or unusable material.

For a *rigorous* test, we need the following:

- **Unambiguous test objectives**, stated so that it is clear whether the materials worked as intended. (See Exhibit 11.1.) In some ways, the objectives are the heart of the test, because they make the "tasks" operational, testable. As Exhibit 11.1 shows, well-made objectives can be assessed by an independent third party.

- A **test protocol**, that is, a research design identifying the subjects to be used, the data to be collected before, during, and after the test, the information products to be tested, and the criteria for acceptance. Typically, the protocol will include interview questions, and, in more sophisticated cases, statistical rules of inference, such as the number of subjects who must complete the test tasks before the material is judged usable.

- **Test materials**, including printed instructions for subjects and any other "handout" material needed for the study. Typically, usability tests of software also require test datasets, hypothetical files that are to be manipulated by the subject. (Usability tests of hardware documents, of course, require that an appropriate version of the hardware be available as well.)

- A **subject sampling plan**, that is, a scheme to ensure that the subjects used in the study are representative of the intended users of the documents. "Accidental" samples are not random samples, and the quality of the inferences drawn from the test is a function of the representativeness of the sampling plan.

- An **unobtrusive test setting**, in which the subjects are free from the influence of the testers. Ideally, the subjects should be alone with the test materials, but, at the same time, the testers should be able to observe them through two-way mirrors or video cameras. It is essential to know that the subjects actually looked at and used the tested documents!

Most of the firms who want rigorous, lab-like usability tests assign the task to a group with appropriate skills, such as human factors psychologists or quality assurance engineers. Increasingly, though, the less formal methods are learned as part of the professional training of a technical writer.

11.2 Shortcuts and Compromises for Usability Tests

Whether laboratory methods or field research is used, high-quality usability research proves expensive. Many firms that start out with ambitious plans decide to compromise on the type or number of subjects and observations. Or even to "fold" the usability study in with other system tests.

Usability testing is expensive. It is possible for a full-scale usability test, with its changes in the product tested, to equal the cost spent on documentation up to that point. Such testing is justified only where the potential risk or benefit is great enough: manuals with huge readerships; systems that perform highly sensitive work; users whose ease and satisfaction are critical to the success of the product.

Perhaps the greatest cost of usability testing stems from its demand for skilled people that many firms do not have on staff: psychologists, psychometricians, social researchers, survey designers. And sometimes it also requires dedicated facilities: labs, video, extra terminals or PCs.

For various reasons, many firms decide that they cannot afford to do usability testing with full science and control. So they cut some corners and make compromises with formal rigor:

Type of Subjects—The subjects in a usability test should be representative users, *not* people involved in the development of the product or publications. But, for reasons of cost or confidentiality, some firms use employees or members of the group as subjects. They try, of course, to pick people who are appropriately naïve, not possessing outside knowledge that would corrupt the test results. But often this selection is too casual, rather like the 1960s tradition of showing drafts to secretaries to be sure they were understandable.

Number of Subjects—Measurement specialists want to be sure that the successes or failures in a usability test are not attributable to mere chance. In practice, many firms wanting to save time and money try the material with only two or three subjects. Some even confine their research to one exhaustively studied subject: a case study. But small samples and case studies, though they sometimes reveal startling truths, are usually unreliable. They give anecdotal insights, but not unambiguous results. Their greatest benefit is that they suggest research questions that need to be addressed more formally.

Design—A maxim of testing is that we must test one thing at a time. For example, if we are testing the manual, the application should be stable, "constant." But many firms, feeling the pressure of time, elect to test their publications as part of the "beta test" of the new product or system. In effect they fold the two tests together. The problem, of course, is that it is often difficult to know where a problem lies.

Any compromise in the plan of a usability test, no matter how worthy the economic motive, has the potential to undermine the integrity of the test. Put bluntly, with the wrong subjects or protocol, the results of the test may be meaningless.

This is more than an academic or theoretical complaint. A large part of the documentation profession has come to equate the term *usable* with *successfully tested in a usability test*. This definition, far more narrow than the view espoused in this book, means that a poorly designed usability test can lead to a misplaced confidence. Fundamental flaws in the documents, such as confusions of audience and function, can be camouflaged by pseudoscientific results. (Consider the many cases of misleading quality-assurance testing as a similar example.)

Of course, there is another view that holds that this preoccupation with scientific rigor is old-fashioned and overdone. In this view the people conducting these shortcut tests are aware of the flaws and have sufficient common sense to temper their conclusions. From this perspective, usability testing is a kind of technical journalism, investigation of a problem augmented with a bit of science. Given resourceful investigators, aware of their biases, we will nearly always be better informed than if we had done no test at all.

Again, much is riding on the choices you make in conducting usability tests, not the least being an unwarranted sense of confidence in a publication that is quite flawed. Before you or your organization embarks on such a project, be sure you can answer these questiions:

- Does anyone on this project have any formal training in social research or tests and measurement?
- Are the people conducting this test sufficiently independent of the developers? Can they be objective and evenhanded? Is it really wise—as is increasingly the practice these days—to entrust the testing to the creator of the object being tested? This has never worked well in programming.
- Is the timing of this test such that there is irresistible pressure for the document in question to pass? Is there time to make important changes if they are needed?

If the answers are unsatisfactory, you might want to reconsider your schedule. Or you might want to engage an independent contractor to do the study for you. This is usually the most economical—and persuasive—approach to usability testing.

11.3 Stereotypes and Traps in Usability Testing

Every sentence and diagram in a manual can be clear and correct—while the book as a whole remains unusable. *That is, the manual could contain the wrong sentences, procedures, and diagrams for the intended audience. Or needed information could be hard to find (though clear once you locate it). Or the manual could be suitable only for its first reading, but inappropriate for later reference.*

With few exceptions, after-the-fact, "one-off" usability tests do not address problems of analysis and design. Nor do they deal with such problems as the interaction between information products (like manuals) and information services (like training).

Usability Testing and Document Overhead

For many readers, the big problem is less one of understanding the instructions than of *finding* the right instructions to read. The issue is **document overhead**: the effort expended by the reader in, first, locating the right starting point and, then, jumping to the consecutive positions in the book. The overhead in a book is directly related to the frequency with which the reader must read something other than the next word or turn to something other than the next page.

Any usability test that tackles one component of the manual at a time—the "unit test fallacy"— will probably overlook the overhead problem. A complete draft has so much structural inertia— and so strong a commitment from its authors— that it makes little sense to raise organizational questions after the fact. When authors find structural flaws in a "finished" draft, they tend to act like programmers who find structural flaws in a finished program: They patch and plug until the problem appears to go away.

Usability after the Neophyte Stage

Certain parts of a manual are read once or twice; others are consulted repeatedly. Although the initial reaction of a reader to a manual may be an important predictor of its subsequent usability, it is hardly the whole story.

Usability tests, in the main, record first impressions. Typically, the subjects are exposed to a text and a task/problem for the first time. Usability is measured mainly in how well this first experience goes. (A small amount of excessive overhead, for example, will scarcely affect the novice, who often *expects* the first trial to be difficult.) But often material that serves well on the first instructional passes becomes clumsy and unresponsive when the more experienced reader consults it for reference. The danger is that usability can degenerate into a cliche, like "user-friendliness," with its unmistakable bias toward ease-of-learning rather than ease-of-extended-use.

Usability testing needs a longitudinal component as well.

Usability versus Maintainability

The most problematical trade-off in developing documents is choosing between usability and maintainability. A usable manual has *conventional page numbers*; many firms use *section numbers* instead (for example, DP0019, 3 of 31), knowing that their publications will be updated so often that page numbering would be a nightmare.

A usable publication *repeats* certain instructions, and even some figures, as a way of reducing the amount of page-flipping and overhead; many firms will brook no repetition, arguing that the problems of maintaining text and exhibits increase exponentially with the repeated appearances of the items. A usable set of manuals *includes some overlap*, so that the user will rarely need to consult two books to perform one task; many firms will not allow any overlap, fearful that the common material will be updated in one volume but not the other.

Most usability testing, though, completely ignores the issue of document maintainability. Is there anything in a typical usability test that tells the developer whether the book tested will be easy to revise, reuse, or cannibalize for later publications?

Usable Books versus Unusable Systems

Some organizations test an early chapter or chunk of a book—"prototyping," they call it. In contrast, by the time most draft documents reach a usability test, the *system* documented will be virtually beyond change. (That is, the politics and economics of the organization will resist any attempt to change the system itself.)

How can a usability test discriminate between certain difficulties attributable to the book and those in the product the book explains? If the manual shows clearly, for example, that a certain transaction is error-prone, who will have the patience and discipline to revise the transaction?

And there are even thornier issues. Is the document *usable enough*, given the extant delays in implementing or shipping the product? Is the manual usable enough, given the priority of a particular product or audience? What are the opportunity costs of continuing to increase the usability of the publication? Would it be smarter to improve the user interface than to improve the user manuals?

Usability tests rarely address essential questions of policy and profit.

12. MAINTENANCE: SUPPORTING AND UPDATING USER DOCUMENTATION

12.1 Maintaining Documents: Stimulus and Response

All documents need changes, from the day they are published. Each impulse to change or revise a document is a stimulus; and the rule governing the correct response is the maintenance standard or policy.

To assure that a manual or set of documents is maintained, you must assign someone to the task. For every document, someone must see that it is distributed correctly and that updates and supplements are sent to the right people at the right time.

All manuals will need to be changed. As Exhibit 12.1 shows, no matter how carefully they are reviewed, your manuals will respond to such stimuli as

- **technical errors**—incorrect or incomplete technical information about the system or product
- **technical changes**—minor modifications in the system, made while you were preparing the documentation, with or without your knowledge
- **communication errors**—ambiguous, unclear, or misleading text and diagrams in your manual; errors of grammar or mechanics
- **system enhancements**—major changes and new features added to the product or system, scheduled or "ad hoc"
- **policy changes**—new rules on what must or may be done, by whom

Although someone must feel responsible for keeping track of these problems, that responsibility need not result in an endless stream of up-to-the-minute bulletins, warnings, and releases.

Despite the documentors' understandable wish to have all manuals current and correct, all manuals are out of date anyway. The question is not "How can the manuals be instantly updated or corrected?" The question is "Which changes can be held for a while—batched—and which must be communicated at once?" Of those that are batched, which can be held for only a few days or weeks? Which for several months?

In fact, there are four main ways to respond to a stimulus:

- **Internal change** is a correction in the master version of the document, that is, the material kept in the files of the person responsible for maintenance; this file contains modifications in the documents, areas that need modification, and release schedules. Everything in this internal file is urgent and should be kept as current as possible.
- **Immediate update** is sending a hot bulletin to every user or document owner; obviously, it is a tactic that should be reserved for important messages. A flurry of emergency bulletins creates confusion and gives the impression either that your system is in chaos, or that you "cry wolf."
- **Batch update** is the collection of several changes in one set, published by the calendar (once-per-month or once-per-quarter) or when the quantity of the material exceeds a certain threshold.
- **New edition** is the ultimate batch update; the documentor incorporates all the modifications since the last edition into a new edition, removing what is obsolete and replacing what has been modified. Then the

users get a new version—ideally, only after they have handed back the old one. (Contrary to what you might expect, a new edition is often more cost-effective than supplements, provided one considers the true costs of inaccurate and incomplete documentation.)

There are, then, several ways to respond and several different levels of urgency (as in all engineering problems). The fact remains, though, that many zealous documentors are too eager and often create more confusion than clarity with their incessant updates.

Exhibit 12.1: The Mediating Role of Maintenance Policies

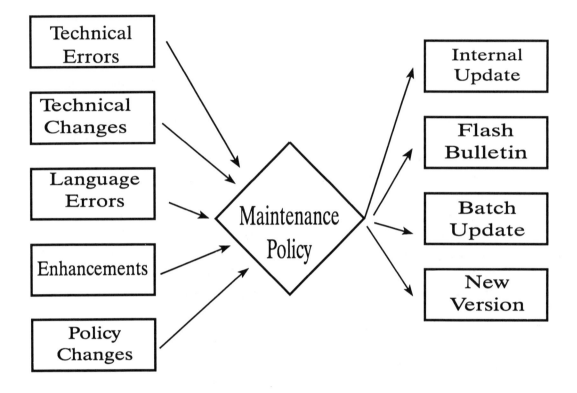

12.2 Information Support for Document Maintenance

For manuals and documents to be maintainable and modifiable, they must have been designed that way. Modular publications, tested when they are still models, not only reduce the need for subsequent changes, they simplify the process of making and controlling whatever changes must be made.

Programs are made maintainable in the early stages of their design; after a program has been coded, it can be maintained, but not made more maintainable. Maintainability is an aggregate measure of the ease and speed with which bugs, flaws, and other inadequacies in a system can be located, defined, and corrected: often the best predictor of its cost-effectiveness.

Similarly, the documents and other information products that accompany systems also must be maintained and modified. The systems change or develop bugs, and the accompanying user documentation must be changed. Or the documents themselves can manifest weaknesses that are independent of the systems they accompany. In time, you will realize that maintaining and modifying publications costs more than writing them—and that publications which resist maintenance are likely to fail, substantially reducing the usability of the systems they support.

The most straightforward way to maintain a user manual is to, first, find the modules that need to be changed and, second, repair or replace them. For modifications, this may also entail finding the right place to add a module and, then, inserting it.

Exhibit 12.2a: Module Profile

```
Module Name: Adding a Record   Module File No.:B-008

First 60 Characters:
    To add a record, type the name of the file in the GOTO win

Superior Modules: B-002 Using the Four File Transactions

Subordinate Modules: B-028 Trying to Add a Record That Already Exists
                     B-029 Trying To Add a Record With Key Data Missing

Descriptors:

    Program/System: DB-3, Real-Estate Manager, Loan-Manager
    User Tasks: file creation, file update, new account, new record
    Audience: end user, realtor, loan officer
    Site/Installation: ABCO Finance, Goldschmidt & Wong Real Estate
    OtherPublications/Products in Which the Module Appears:
                G-3, G-4, G-5; F-1, F-2, F-7; R-1, R-5
```

When user documentation is modular and structured, you can maintain a directory of all the modules, coded so as to define the systems, topics, applications, installations, or other descriptions that are relevant. Such a scheme allows you to search the file for all the modules affected by a particular system change. And also allows you to generate new documents from old modules.

If you view documents as unique sets of modules, you can maintain a directory like that illustrated in Exhibit 12.2a. Because it is likely that a module will appear in more than one place, such a directory tells you all the publications that are affected by a particular change in the system.

In large and sophisticated organizations there may be alternative versions of the same technical content expressed in **equivalent modules**; the directory illustrated in Exhibit 12.2b allows the documentor to map all the consequences of a technical change onto the various publications that need to be changed.

There are other, simpler anticipatory design choices that can make your documents more maintainable. For example, manuals in loose-leaf binders, obviously, are more agreeable to change than bound books. One-page modules, printed on one side of the paper, are the easiest to add, remove, and insert. And they are probably the best form of module for documents that need to be changed continuously. On the other hand, though, they tend to make many manuals choppy and filled with complicated references and loops. The maintenance advantages of the one-page module—with text and exhibits on the same page—may have to be traded-off for the usability advantages of the two-page module.

Exhibit 12.2b: Directory of Modules

Module	Equivalent Modules	Publications/Products
B-008	R-006, D-120	G-3, G-4, G-5, F-1, F-2, F-7, R-1, R-5
B-028	R-061, D-121	G-3, G-4, G-5, F-7, R-5
B-029	R-062, D-122	G-3, G-4, G-5, F-7
C-110		G-2, G-3, G-4, G-5
C-115	R-090	G-4, G-5, R-5
C-240	D-600	G-3, G-4, G-5, F-7, R-4, R-5

12.3 The Maintenance Paradox: The More the Messier

Without a thoughtful policy for the maintenance of documents, there tends to be a random or haphazard distribution of supplements, bulletins, releases, and updates. What many do not realize is that each supplement to a manual can actually double the number of alternative versions in circulation. And only one of these versions is correct.

Again, it is a truism that all manuals are out of date and that they contain at least a few errors. This is no more remarkable or deniable than the claim that all complicated programs or devices have bugs—including some that have not yet been recognized.

It is also a truism that all user lists, distribution lists, and route lists contain errors. And the longer the list, the more inaccurate and out of date it is. That is, any attempt to communicate bulletins and changes to all the people who are using a certain document—or a certain system—will be frustrated by the inaccuracy of that list.

These first two truisms—that all documents contain errors and that all distribution lists are inaccurate—are almost natural laws of technical communication. If we add another law, the Second Law of Thermodynamics (the entropy principle), it becomes more understandable why attempts to update and revise publications so often fail.

Not only is there a continuing struggle to recognize and write up the needed changes; not only is there an eternally frustrating attempt to identify all the people and places that should receive the supplements and updates; there is also a vast set of random and perverse forces that conspire to misdirect and distort the effort. Mail systems, private or public, make errors—even if they are electronic or fax systems. Also, the recipients tend to misplace, misapply, misread, and otherwise abuse the bulletins. In how many manuals, for example, are all the supplements

still wrapped in clear plastic, waiting to be incorporated?

The net effect is that every supplement to a manual—even though its purpose is to produce current, consistent documentation—may double the number of versions in circulation. (Some people receive the supplements; some don't.) When the original manual appears, there is only one version in circulation. (Not counting, of course, the unofficial versions extracted and created by industrious users.) With each added supplement, the number of alternatives doubles.

Thus, two supplements yield four versions, and four supplements yield 16. After 10 supplements there could be 1K versions: 1024!

This discussion is not intended to be humorous. Anyone who has tried to distribute corrections and updates to a large set of operators or customers knows that every possible misuse, misplacement, and mismanagement of the documents will, in fact, occur. Incorrectly addressed materials disappear; correctly addressed materials are nevertheless mislaid. Materials that supersede older versions are stored in a desk, while the obsolete pages remain in force at the terminal. Often, one cannot find two identical versions of an important publication.

Although nothing can prevent completely this proliferation of misinformation, several measures can ameliorate it:

- **Limiting the number of supplements and releases**, keeping them in large batches, will reduce the noise in the documentation channel. Releasing these batches on a regular schedule solves another serious problem as well; it lets the users be confident that they have received all the supplements. When books are updated irregularly, the user is never sure.

- **Putting as much documentation as possible into the system itself**—thereby reducing the quantity of obsolete "hard copy"—will contain the problem.
- **Limiting the updates to a single, authorized source** will reduce confusion and resolve conflicts.
- **Requiring technical specialists to review and approve the documentation** *before* **it is sent** will reduce the quantity of updates and corrections, especially the corrections of the corrections.

If you consider the true costs of misinformation, you will realize that responding to problems caused by the incorrect instructions costs much more than writing and publishing better documents. Lost work, inexact work, operators' downtime, emergency visits to troubled sites, additional consulting, training, travel, re-entry of data, revising of documents—all these can conspire to make the issuing of a whole new edition less "expensive" than a two-page supplement.

12.4 Can Old Manuals Be "Modularized"?

Few documents start from scratch; usually there is an old tome to be incorporated or updated. On occasion, an old book can be recast into a structured format, but the change may be a little more than cosmetic, not really providing the benefits of a brand new, tested, modular manual.

Often, the assignment for newly hired documentors is to finish, update, upgrade, revise, or otherwise resuscitate some unacceptable publications. They are asked to begin long after the time when most of the document design decisions should have been made.

What about these existing documents? Can documentors charged with the task of editing or revising old manuals make use of the structured approach? Can an inaccessible, unreliable manual be made more usable?

Perhaps. The editors at Hughes Aircraft, when they first publicized their method of modular publication—the STOP (Sequential Thematic Organization of Publications) technique— reported that they were able to recast old documents into the new two-page format. Partly, their success was due to the nature of the publications they were working with, many of which were already equal mixes of text and exhibits.

Given the right publication, the process can be almost fun. All the pages of the existing document are laid end-to-end, and the team of designers goes through the text and pictures marking off module-sized chunks of material, writing new headings or headlines, occasionally—but rarely—even rearranging the sections, or moving an exhibit from the appendix to the text.

Whether this is a good idea depends on several factors. As already mentioned, some publications, which are closer to the structured

Exhibit 12.4: Retrofitting the Unstructured Text

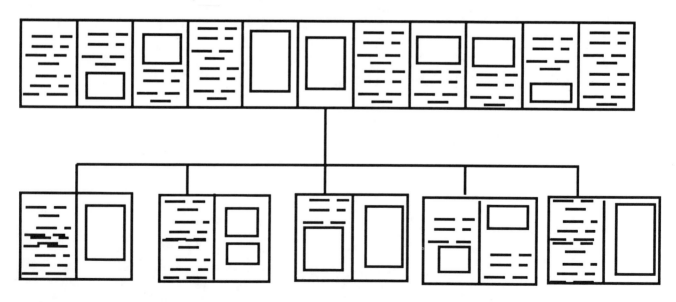

format than others, call for less complicated reworking. Some manuals, because they are filled with good writing and pictures, are especially worth saving, and they justify the effort. But many documents (like many old programs) are obsolete and clumsy. Trying to save these is little more than yielding to the too common myth that it saves time and money to reuse existing material instead of creating new things.

Again, some old manuals lend themselves quite easily to "modularization." I have seen people "back out" a storyboard from an existing manual that never had one in the first place. In many cases, though, the effort to recast and retrofit an old publication is greater than the effort to generate a brand new one.

But, even when an old book lends itself to recasting in the more readable modular format, please remember that modularity is not just an attractive way of presenting material. It is not just cosmetic.

Modular design, whatever its aesthetic benefits, is also a way of assuring that the documents are maintainable and modifiable. A modular format imposed after the fact may improve appearance, but it might not significantly improve the maintainability and reliability of the book that has been revised.

I agree with Yourdon and Constantine in their discussion of recasting old computer programs:

> It is all but impossible to simplify significantly the structure of an existing program or system through after the fact modularization. Once reduced to code, the structural complexity of a system is essentially fixed. It is thus clear that simple structures must be designed that way from the beginning.

—*Structured Design*
(Englewood Cliffs, NJ: Prentice-Hall), 1979, p. 35

Recast and retrofit old manuals if you like. But do not miss the opportunity to engineer a publication from its inception.

PART 3

Online Documentation and Internal Support

13. USER DOCUMENTATION WITHOUT BOOKS

13.1 The Full Meaning of "User Support"

Into the early 1980s, most developers believed it natural and inevitable that customers, users, and field installers and service people would need support. Moreover, the natural way of providing it was through publications, typically large, difficult, and demanding. The newer view, though, is that support may be a euphemism: that what traditional manuals try to do is ameliorate the defects in the system.

Virtually all users of computer and communication products presume that these products will come with documentation. They expect mainframes and giant telephone products to come with libraries. Indeed, for sophisticated PC software, they not only expect the product to come with immense publications, they also expect to find hefty third-party treatises at the local bookstore.

For any complicated product there are support needs: initial setup, orientation, and training; responding to bugs and anomalies as they inevitably appear; customizing and adapting the product to fit the peculiar circumstances and equipment of the customer; getting out of user-created messes; recovering from crashes, power spikes, and other acts of God; resolving jurisdictional disputes when one program wars for RAM with another.

The issue is whether this litany of painfully familiar "support" cases is necessary. Granted, there will always be unanticipated problems caused by variations from machine-to-machine. Still, is it possible to develop technology, espe-

Exhibit 13.1: Mismatch between User and System Mentalities

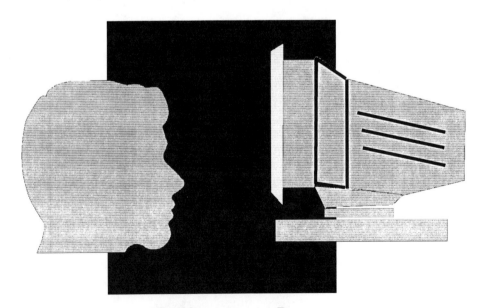

Mismatch

cially business technology, that *doesn't need so much support?*

With the right interface and introductory training, could we not eliminate much of the elaborate procedural discourse from our manuals and encourage users to explore the system instead? That is, for a comprehensive manual substitute John Carroll's *minimal* manual (*The Nurnberg Funnel*, Chapter 6).

An uncharitable view of user support—the way it is seen by many human factors consultants, for example—is as a set of materials and procedures that compensate for the inadequacies of design, especially in the user interface. In large manufacturing firms, for example, we often see a clash between the engineers/programmers, who think it natural for new products to be complicated and esoteric, and human factors psychologists, who are not so persuaded.

Whether or not systems *must* be hard to learn and error-prone, it is unmistakably true that most of the writers and trainers at work right now are, in fact, filling in the gaps between the profile of the system and the profile of the user. It has been the natural role of such people, since the 1960s, to teach arcane vocabularies, demonstrate tricky procedures, and generally enable users and customers to adapt their natural ways of thinking and working to the demands of the machine. Indeed, a few large firms have built their success on a reputation for support, just as more than a few vendors have made profits from teaching customers, for a fee, to use the products they already paid for.

Even though it may be impossible to eliminate all external support from sophisticated systems, the current view is that *most systems need far more support than they should.* That is, users are driven to hotlines by a poor installation routine; they are shunted to manuals by needlessly cryptic screen messages or obtuse command syntax; they are compelled to use Help screens by free-form procedures and indecipherable menu choices. Or, put another way, a well-made system needs a relatively small support envelope, while a poorly made system will need to be bundled with a complicated entourage of information goods and service, including, in some cases, ponderous libraries that nearly everyone loathes.

The problem is partly in the tradition. To solve it, we need some new attitudes:

1. **The time of the user is more valuable than the memory resources of the system**; Users should rarely have to decode messages that could have been stored, English-like, in the system.

2. **Systems have no right to scold users for making mistakes.** In fact, the mistake is really the system's failure to understand the user. Moreover, blame is irrelevant; recovery is what we need.

3. **Messages that appear on the screen— especially those containing instructions— should be written and edited by people who can write well.** Most coders do not meet that requirement.

13.2 Breaking the Grip of Manuals

There are six problems that increase our grudging dependence on user manuals: free-form procedures, cryptic menus, undecipherable prompts/labels, unintelligible system/ error messages, error-prone data entry, and unhelpful Help screens.

Free-form Procedures and Routines

The most fundamental, continuing problem is the procedure that places its entire burden on the user. In its most extreme form it is the familiar blank-screen-with-blinking-cursor or solitary DOS prompt. In its milder form it's any command-driven system that requires users to re-member names, definitions, syntax, limits, and so on.

The most dramatic way to weaken the grip of manuals is to use pointing and recognizing in place of recalling and typing. It is the essence of what has come to be called "friendliness," and it is also the main weapon in breaking our depend-ence on publications.

Cryptic and Ambiguous Menus and Options

Ironically, although menu-driven choices are the principal way of eliminating our dependence on documents, bad menus are almost as serious a problem.

The greatest weakness in much online infor-mation is *plain old bad writing*. Why should a menu say "Execute a Format" when it means "Make a Chart"? Which of two options should I select: "Store Chart" or "Store Composed Chart"? (And what if I told you that the wrong choice could undo hours or days of work?)

Why do some menu writers believe that menus should contain only one-word entries? And how do they expect me to choose between "Copy" and "Replicate"? If I want to change the size of a font, why do I have to choose "De-faults" from the menu?

Undecipherable Prompts/Field Names

Because many programmers learned their craft in an era when machine resources were dear (and good programmers didn't waste them), our screens are still filled with compressed and abbreviated words, starting with "usrid" and getting worse.

Even worse than the single-word prompts are the unreadable clauses and sentences, often (it appears) written by people with only a passing familiarity with English. What is a normal person to make of "Hitting the Space Bar Unselects Your Selection"? How can users tolerate "Press Enter to Exit"?

Generally, it is unproductive and uncaring to expect people to read messages they cannot unpuzzle. Especially when rudimentary copy-editing would fix most of the trouble.

Unintelligible System/Error Messages

Every so often, a system seems unable to do what the user wants. In these cases, the system tries to communicate its difficulties to the users. If the problem is in the system, it sends a system message; if the problem is with the user's ac-tions, it sends an error message.

The trouble with these messages is their *tradition*. Traditionally, error messages were coded or encrypted. It was reasonable to tell the user "Error Code 11-11 has occurred at Line 15050."

Also part of the tradition is the attitude that any failure of the system to understand the user should be regarded as the user's error—as though it was the user's fault that the system is so compulsive about spelling and syntax.

System and error messages (and, by the way, it's probably time to retire the phrase "error message") have only one purpose: *to get the user to the next step of processing.* What is a speaker of normal English to make of

> No application is installed for this type of document. To run a particular application whenever you open this type of document, select the application and install it.

Error-Prone Data Entry Mechanisms

One way to mediate the effect of poor error messages is to eliminate errors. Wherever possible, users should be able to select what they want from a menu, and activate that choice with only a keystroke or two. (Those who want to type should have the option.)

Moreover, the user interface should work as consistently as possible. "Modes," which change the functions of keys, can also drive people to their manuals. When the same operating system wants the <BREAK> key in one mode and the <RETURN> key in another—when both seem to be doing the same job—the manual writer had better prepare.

Unhelpful Help Screens

Help screens are an attractive alternative to reference manuals. With few exceptions, users do prefer to look up codes and jog their memories by pressing a "hot key" that activates a germane Help screen.

But Help screens must be designed for readability. Fewer than half the cells in the grid should have characters in them; a screen full of prose paragraphs is even harder to read than a page full of the same.

And, finally, Help screens must also be well written. If anything, given the difficulty of reading messages from a video screen, they must be better written and edited than traditional publications. Under no circumstances should they by written by people who can scarcely write.

13.3 Some Relevant Principles of Human Factors

Designing a user interface is a "human factors" problem. (Not just a programming problem; not just a writing problem.) The 1980s proved that three principles are paramount: Pointing is better than typing; recognizing is better than recalling; selecting is better than reproducing.

A high proportion of the errors made by users are nothing more than typing errors. A simple transposition will provoke the system to such irksome responses as "file not found." With some exceptions, *the longer the string of characters to be* *typed, the greater the opportunity for error.* (Unless the longer string is more English-like and, thus, easier to remember.) Wherever possible, the user should be enabled to point (using

Exhibit 13.3a: Pointing vs. Typing

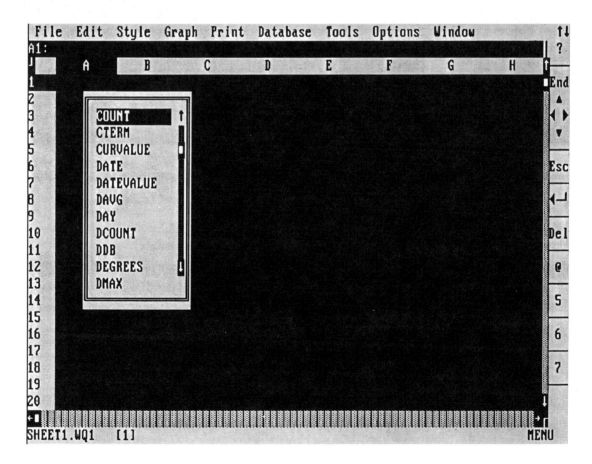

the cursor keys, a mouse, or some other pointing device) at the choice on a list or menu.

Generally, the fewer keystrokes demanded of the user, the fewer problems and mistakes. Whenever possible, users should select (by "pointing and shooting," as they say) from the list of options. They should not be asked to reproduce a word or string of characters on the prompt line. Indeed, they should not even be asked to type the letters and numbers of the choices on the menu.

To a large extent, the use of selectable options—menu-driven and prompt-driven programs—has become a working definition for **user-friendliness** or **usability**. And although there are times when it is inappropriate, it is still the best way to build an interface and *reduce the need for user documentation*.

Similarly, users should not be asked to remember things, or to look up things in ponderous manuals, when it is so much easier to recognize things from a list. When users want to know their options, the table of allowable entries should be ever-present on the screen or immediately available through pull-down/pop-up menus, or through a traditional Help facility.

Even the presumably simple task of typing a letter or number to select a menu item is less reliable—more error-prone—than allowing the users to point at and select their choices.

Exhibit 13.3b: Pull-Down Menus as Recall Aids

14. STRATEGIES FOR ONLINE DOCUMENTATION

14.1 Online Documentation—Five Fronts

In its weakest form, online documentation is merely reading traditional user documents on a screen or other display. In its strongest form, it is incorporating teaching and reference material into the system itself, as seamlessly as possible. There are five ways to approach online documentation: books on a disk, Help screens, computer-based training, improved user interfaces, and hypertext.

Like other fashionable expressions from computer technology, "online documentation" has many meanings. It covers a range of support options, everything from shipping old-fashioned documentation on floppy disks, to redesigning a system so that it needs less paper documentation. For some people, online documentation means reducing a computer to a kind of expensive microfilm reader, with page after page of dense text (often without graphics).

All forms of online documentation, from the least to the most ambitious, are legitimate and helpful. And, even though there are some forms of support that do not lend themselves to online format, every system can benefit from having more or better online support than it has now.

There are five fronts for approaching online documentation, which can exist in any combination:

Books on a Disk are traditional publications that have been read into an electronic file so that the pages can be read on the computer screen. In the simplest form, the table of contents and index act as menus, from which users select the passages they need to read. In more complex approaches, the document file has a retrieval utility—a "lookup engine"—that permits users to search by keywords and phrases.

Online Help is the practice of creating screens that instruct or answer questions and attaching them to particular elements in the master or application system. In its simplest form, the call for Help takes the user to the menu of a book on a disk; in more sophisticated approaches, the system senses the context of request—the particular screen or field that caused the impasse—and presents one or more screens aimed at that special need.

Computer-based training uses the power of the computer to present, manage, or evaluate instruction. The more sophisticated the CBT program, the more interaction it demands of the user: asking questions, assigning tasks, branching as a function of the user's responses.

Better user interfaces means eliminating the several factors that drive people to manuals (and Help screens) by simplifying and clarifying the way systems communicate to users and, thereby, reducing the chance that users will commit "errors." It entails everything from replacing command-driven programs with menu-driven ones, to rewriting menu language for clarity, to creating "shells" or "environments" that allow easy manipulation between screens, or even across applications.

Hypertext is a relatively new programming approach that can be used to implement any of the four fronts above, or to supersede them. Hypertext, the linking of screens/files so that readers can navigate through a web of topics (according to their unique interests) not only allows people who used to be writers to create tutorials, Help screens, or interactive books without the aid of programmers, but also permits nonprogrammers to link these elements into new applications of their own.

Exhibit 14.1: Table of Online Fronts

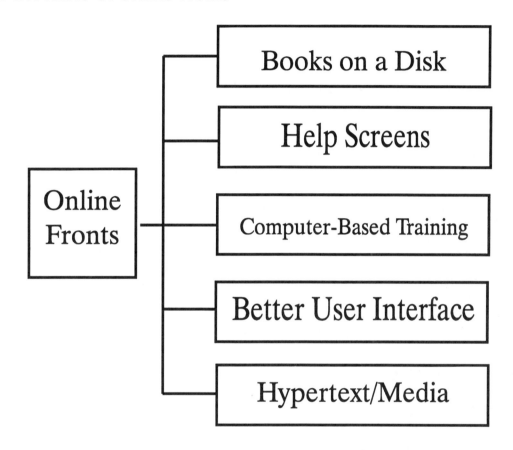

14.2 Books on a Disk

The easiest way to create online documentation is to convert the book or library to an electronic file and allow it to be read on a screen. In its least ambitious form, the table of contents or alphabetical index becomes a menu. In more ambitious forms, the stored book is accompanied by a "lookup engine," a utility that searches by keyword or phrase. Although such systems may yield nearly unreadable screens, they are still an important advance on the traditional document library.

The term *online documentation* has changed meanings with the changes in technology. As recently as the 1970s, writers who claimed to have "put their documentation on line" sometimes meant merely that their entire manual (but not the graphics) was done on a word processor. In earlier days some firms even referred to the practice of shipping their publications on disk—to be printed by the customer—as online documentation.

Even today, though, there are sophisticated firms to whom online documentation does *not* refer to better menus and Help screens but to the process of converting an entire library of technical publications to electronic form. In effect, this electronic publication is separate from the application system, meant to be read separately, in much the way that a technician will "stop working" and consult a manual. Particularly companies that manufacture computers and

Exhibit 14.2a: A Cluttered Screen

```
↓↑ PgDn PgUp End Home  F=Find Esc=Exit ?=Help |
           program to keep logs of time spent on client projects, calculate the
           amounts to be charged based on hourly rates you specify, and print
           monthly records of time spent on the projects.
515        AS-EASY-AS v3.01 - AS-EASY-AS is as close as your going to get to
           Lotus spreadsheets.  A well made spreadsheet program that rates and
           compares to the expensive spreadsheets.  Some of the options that are
           avaible to you are  2048 rows by 256 columns, POPUP or PANEL menus (You)
           decide which is best for you), powerful graphics that will let you
           create BAR, STACKED BAR, LINE, XY ,PIE, HILO, SEMI-LOG and LOG-LOG
           graphs on your Epson compatible printer , over 50 {MACRO} programming
           commands, over 43 @FUNCTIONS for math, statistics and finance, 3D-
           SIMULATION, you can reallocate spreadsheet memory upto 100 planes.  HELP
           screens , built right into the program , over 50 pages, and if that
           wasn't enough AS-EASY-AS is highly Lotus compatible.  UPDATE
518        PIVOT v1.01 - AS-EASY-AS one of the better spreadsheet programs just
           got better.  Now you can control your printer setup more easily with
           Piviot. PIVOT! allows you to print your AS-EASY-AS worksheets in either
           portrait or landscape orientation and in the font of your choice. (Two
           intenral fonts currently available). The GCHAR (Graphics CHARacters)
           program, distributed as part of this package, is used to develop your
           own fonts, which can be accesed by the PIVOT! program.  All in all, the
           two programs form a powerful combination which is an invaluable
           companion to AS-EASY-AS and other compatible spreadsheets.  Requires
```

telephone equipment are inclined to reduce the several thousands of pages associated with a mature machine or a huge switch to a miraculously small compact disk that can be read through a PC.

Such conversions are often artless, consisting in the straightforward reduction of dense, unusable texts into what ultimately become denser and equally unusable screens. (This process converts the computer to a glorified microfilm reader.) Everything that was wrong with the pages—from substance to layout—is now worse. (See Exhibit 14.2a.)

The usability research is clear on this subject: computer screens need even more "white space" than paper. Paragraphs that are nearly unreadable on paper become entirely unreadable on the screen. Such practices as justifying monospace characters on a character-based screen produce panels that even the most intrepid user cannot abide.

But even the crudest book on a disk can be a significant improvement in user support, especially for those situations where there is usually

only one set of user documents. PC users, who typically have all their own manuals, forget that mainframe and mini users typically work in settings where the firm owns only one library of manuals. And this library, moreover, is often at another location! In contrast with this tradition, having one's own library nearby (even in a car or at a remote location) is an important advance, even if the screens do not meet the standards for modern design.

Today, moreover, not all books on a disk are crude and artless. Increasingly, the pages are redesigned as suitable screen panels; by the end of the century, better computer screens will permit more readable characters and screen "typography." And, most important, many modern examples come with "lookup engines," simple search utilities that allow the reader to request particular topics or to type keywords; depending on the product, the system will either jump to a passage containing the keyword or, more often, list a menu of articles or sections addressing the requested topic.

Exhibit 14.2b: Lookup Engine that Searches for Topics

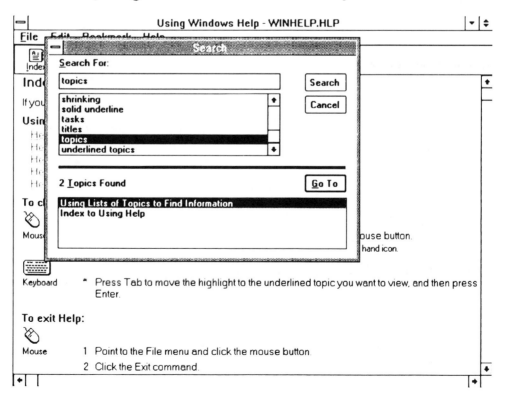

14.3 Styles of Online Help

Every day, developers think of new ways to provide online Help. The main styles are menu-driven manuals, *in which the table of contents becomes a menu;* context-sensitive, *in which the system guesses the user's problem by noting the location of the cursor; and* continuous, *in which certain zones of the screen are automatically filled with Help information.*

Nowadays, almost every sophisticated application has some form of online Help. It is hard to imagine a user support envelope with no plan for this essential component.

But the decision to incorporate online Help does not, in itself, define the scope or approach. Traditionally, there are at least three very differ-

Exhibit 14.3a: Table-of-Contents Menu

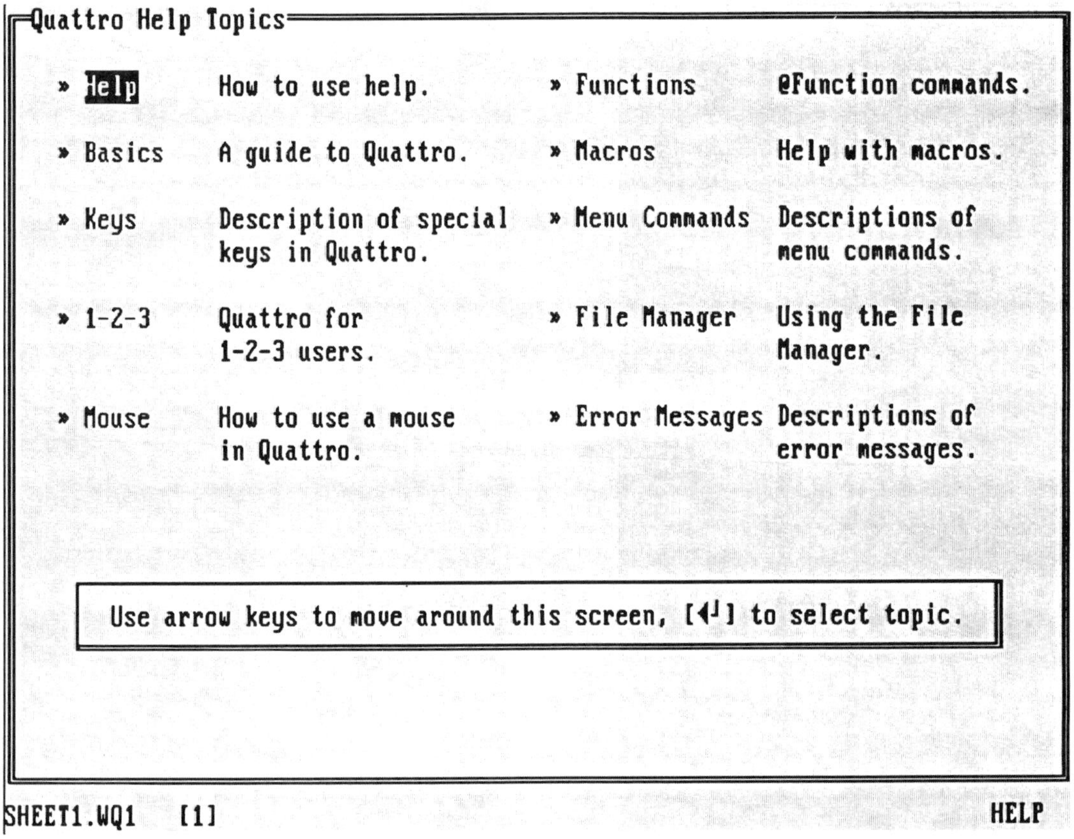

```
┌Quattro Help Topics═══════════════════════════════════════════════════╗
║                                                                       ║
║   » Help      How to use help.      » Functions    @Function commands.║
║                                                                       ║
║   » Basics    A guide to Quattro.   » Macros       Help with macros.  ║
║                                                                       ║
║   » Keys      Description of special » Menu Commands Descriptions of   ║
║               keys in Quattro.                      menu commands.     ║
║                                                                       ║
║   » 1-2-3     Quattro for           » File Manager Using the File     ║
║               1-2-3 users.                         Manager.           ║
║                                                                       ║
║   » Mouse     How to use a mouse    » Error Messages Descriptions of  ║
║               in Quattro.                           error messages.    ║
║                                                                       ║
║                                                                       ║
║     ┌───────────────────────────────────────────────────────────┐    ║
║     │ Use arrow keys to move around this screen, [◄┘] to select topic. │
║     └───────────────────────────────────────────────────────────┘    ║
║                                                                       ║
║                                                                       ║
╚═══════════════════════════════════════════════════════════════════════╝
SHEET1.WQ1   [1]                                                   HELP
```

ent ways to design a Help facility, each with different costs and effectiveness.

The least demanding approach is **the online manual**, a manual attached to the system as an electronic file, rather than as a paper document. The important difference is that the table of contents functions as a menu, through which the user reaches appropriate topics or sections.

This sort of Help component is "unintelligent"; it does not deduce the user's problem. The burden is still upon the user to locate the material in the online "book." Although unsophisticated, it still has certain advantages. First, for those systems with huge documentation libraries, the online book solves the problem of the single copy. In effect, all users have their own copy of the documentation and do not have to leave their terminals to visit a library. Second, for those users intimidated by large books, the online manual is less of a barrier, especially for those who tend to get lost in complicated tomes.

Most users, though, prefer the second style: **context-sensitive** Help, in which the system deduces what the user was doing at the moment of impasse and answers with suitable reference or procedural information.

In the '90s, most Help screens will be tied to particular fields, or panels, or transactions, activated by a "hot button" (usually <F1>).

Exhibit 14.3b: Context-Sensitive Help

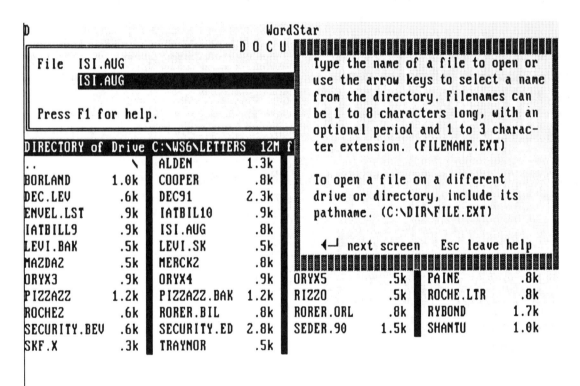

14.3.1 Attaching a Menu-Driven Manual

A fast way to implement online Help is to copy a book of reference materials to a file and arrange to have the Help "hot button" bring up the table of contents for that file. Typically, all calls for Help will result in the same menu, no matter what task the user was performing at the time. As with the book on a disk, the burden is on the user to pick the appropriate chapter or page to study.

The book on a disk is a separate program, run independently from the application. If the same book is accessible from *within* the application—by pushing a "hot button"—then the style may be better described as a **menu-driven** Help utility.

In this mode, typically, any request for Help results in the same screen: a table of contents. (See Exhibit 14.3.1a.) If you know what you need, you will be able to find the appropriate "page."

Other examples are more curious. Exhibit 14.3.1b, for example, is a table of contents with one entry per page. But note the absurd choices ("Options, continued").

Most menu-driven Help utilities use this table-of-contents approach; the menu is organized logically and chronologically. In other examples, though, the topics are arranged alphabetically, like an index. (This approach assumes you have a keyword or phrase in mind.)

Exhibit 14.3.1a: Table of Contents for Help

```
              CALENDAR CREATOR PLUS 4.0
                  Table of Contents

      A. Introduction .............................. 1

      B. Main Menu ................................. 2

      C. Create/Edit Event List ................... 5

      D. Copy/Merge Event Lists ................... 18

      E. Printing ................................. 19

      F. Printer Defaults ........................ 28

      G. Defaults ................................. 33

      H. Converting Old Versions ................. 35




        Enter desired page number    and press Enter.

  PgDn Next Page                      End Index   ESC Return
```

There are, of course, many variations on the menu-driven manual. Some applications incorporate a bit of context sensitivity and show only a portion of the table of contents—a deduction based on the activity of the user just before Help was asked for.

Still other systems hide much of the book and its table of contents. The user asks for Help on the prompt line, specifying a topic (for example, ? ASCII). The Help utility either presents a short menu of likely options or goes directly to the "ASCII" section of the manual.

Exhibit 14.3.1b: Unusual Use of Table of Contents

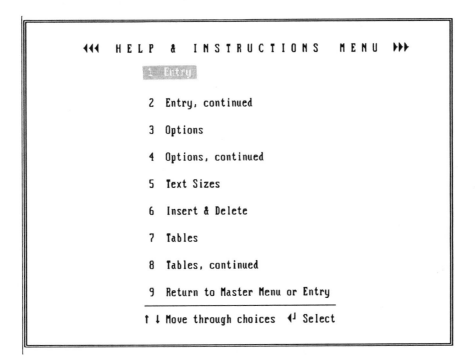

Exhibit 14.3.1c: Alphabetical Index of Help Topics

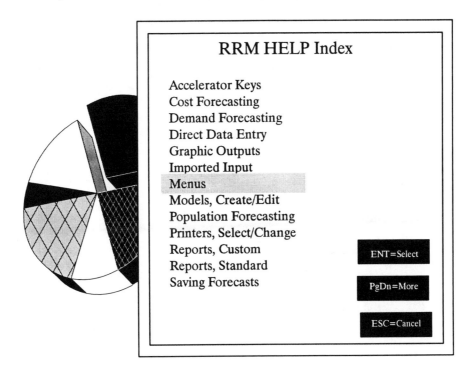

14.3.2 Deducing the Problem with Context-Sensitive Help

The most preferred form of Help is context-sensitive. In this approach the particular Help screen or file is tied to the specific point at which the user pressed the "hot button." In its most sophisticated form, there is a unique Help screen keyed to each application field on the screen, or even to each coordinate on the application screen.

Today's users expect their Help facilities to be somewhat intelligent. That is, the system should be able to guess what problem caused the user to ask for Help.

Context-sensitive Help refers to utilities in which the software contains an elaborate table linking particular Help panels to particular fields, screens, or screen coordinates. In the most typical case, the user is unable to fill in the required answer to a prompt or the acceptable data for a field, asks for Help, and receives either a small table of allowable values (and their meanings) or a brief procedure for responding. Exhibit 14.3.2 shows the most commonplace form of context-sensitive Help.

The context may be broad or narrow. In some systems, there is one Help panel for a whole *transaction* or series of screens; in others, each application screen has a unique Help screen. But the most likely approach, these days, is to have a Help screen for *each application field* or user response.

Some applications limit users to one Help screen per request. That is, there is only one Help panel associated with the particular field. This burdens the developer, who must make sure that the Help screen is the right one for all users. In other systems, though, users may ask for *more* Help. If the first panel does not resolve their problem, they may ask for a second, or even a third. Ordinarily, the first panel is reference, the second panel is procedural, and the third is more discursive, providing general information and background about the system. (Generally, people who need the second and third Help panel are undertrained.)

The main key to success in writing context-sensitive Help screens is to know what the typical user will want for each screen, and to provide *only* that. Help screens that are cluttered with irrelevant discourse can often, ironically, drive users to their manuals.

Exhibit 14.3.2: Field Tied to Table of Acceptable Values

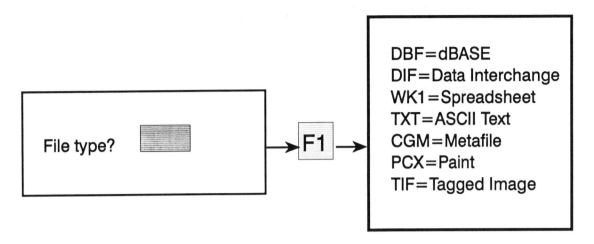

14.3.3 Providing Continuous Help and Prompt Zones

Most Help is voluntary; users ask for it. Some Help is compulsory. Either it continuously fills some part of the screen. Or it pops up in "prompt boxes" tied to particular fields or elements in the screen. In both cases, such Help can occupy a large portion of the screen, reducing the area left for the application.

In some applications, a form of involuntary Help screen is available at all times. That is, a portion of each screen is allocated to a screen- or field-sensitive Help panel.

Example 14.3.3a, for instance, shows how WordStar can be configured to allocate about half of its application screen to a Help table containing a list of the main editing commands.

The advantage of this approach is clear: The inexperienced user—or the user accustomed to some other word processor—does not have to search for the basic skills needed to operate the

Exhibit 14.3.3a: Half-Screenful of Continuous Help

```
 C:PIZZAZZ       P1  L31 C1    .00" Insert Align
 ╔═════════════════════════ E D I T   M E N U ═════════════════════╗
 ║   CURSOR        SCROLL       DELETE      OTHER            MENUS   ║
 ║ ^E up         ^W up        ^G char    ^J help         ^O onscreen format ║
 ║ ^X down       ^Z down      ^T word    ^I tab          ^K block & save    ║
 ║ ^S left       ^R screen up ^Y line    ^V turn insert off ^P print controls ║
 ║ ^D right      ^C screen    Del char   ^B align paragraph ^Q quick functions ║
 ║ ^A word left    down       ^U undo    ^N split the line  Esc shorthand     ║
 ║ ^F word right                         ^L find/replace again                ║
 ╚═════════════════════════════════════════════════════════════════╝
 L----!----!----!----!----!----!----!----!----!----!----R

I bought the upgrade, and even paid $10 for swift delivery.              <
                                                                         <
Imagine my disappointment to learn that the advertised preview feature was not
available on my 286/VGA system; nothing in the brochure indicated that this new+
feature was limited to PS/2 machines.  (And, one must ask, Why?)
                                                                         <
I called technical assistance and they could do no more than tell me to print t+
color equivalence charts.
                                                                         <
Again, I am disappointed.  Is there some trick or "workaround" that will solve +
problem.                                                                 <
 Display Center  ChkRest ChkWord Del Blk HideBlk MoveBlk CopyBlk Beg Blk1End Blk
 1Help  2Undo   3Undrlin4Bold   5DelLine6DelWord7Align  8Ruler  9End-Lin0Done
```

program, nor even ask for Help. The disadvantage is also clear: Much of the usable application screen has been sacrificed to the Help materials. Thus, with such approaches, users must be allowed to change the configuration and eliminate the continuous Help when they no longer need it.

Another widely known example of involuntary Help can be seen in the Norton Utilities. When users choose the various utilities from the interface the functions and switches for each option appear automatically in a kind of "prompt zone" (or "split screen" or "window") attached to the option. As the cursor moves through the list of choices, the prompt zone information changes automatically, providing information that might usually be part of a Help screen.

(Another way of classifying the "prompt zone" is to consider it a secondary menu, that is, a window that amplifies the meaning of each menu option.)

Clearly, the use of prompt zones is less an example of "external" support than of designing a user interface that needs less support. The prompt zone (like the secondary menu) so reduces the memory burden on the user that most traditional and online documentation becomes irrelevant.

Exhibit 14.3.3b: Split-Screen Help

```
╔═══════════════════════════════════════════════════════════════════╗
║  ┌──────────────────────┬═ The Norton Integrator ═┐                ║
║  │ BE  Batch Enhancer    │                          │               ║
║  │ DI  Disk Information   │ LP       LP filespec [where-to-print] [switches] │
║  │ DS  Directory Sort     │    Print text files with a variety of    │
║  │ DT  Disk Test          │    formatting options.                   │
║  │ FA  File Attributes    │                                          │
║  │ FD  File Date/Time     │ LP help.txt help.prn  /N /L4            │
║  │ FF  File Find          │    Prints the file help.txt to the file help.prn, │
║  │ FI  File Info          │    numbers each line (/N), and sets a four- │
║  │ FR  Format Recover     │    character left margin (/L4).           │
║  │ FS  File Size          │                                          │
║  │ LD  List Directories   │ Switch          Default │ Switch          Default │
║  │ LP  Line Print         │                                          │
║  │ NCC Control Center     │ /N   Line numbers     off│ /Pn  Page start #    1 │
║  │ NCD Norton CD          │ /Tn  Top margin       3  │ /Sn  line Spacing    1 │
║  │ NDD Disk Doctor        │ /Bn  Bottom margin    5  │ /80  80-col print    on│
║  │ NU  Norton Utility     │ /Ln  Left margin      5  │ /132 Condensed mode off│
║  │ QU  Quick UnErase      │ /Rn  Right margin     5  │ /WS  WordStar files off│
║  │ SD  Speed Disk         │ /Hn  page Height 66 lines│ /HEADERn Level       1 │
║  │ SF  Safe Format        │ /Wn  page Width   85 cols│ /EBCDIC  Code option off│
║  │ SI  System Information │                                          │
║  │              more...   │ /SET:filespec   File of Lotus-style setup strings │
║  ├──────────────────────┴──────────────────────────┤               ║
║  │ LP                                                                ║
║  └─────────────────────────────────═ Press F1 for Help ═┘           ║
╚═══════════════════════════════════════════════════════════════════╝
```

14.3.4 How Help Screens Fail

The two main requirements for a well-made Help screen are, first, functional cohesiveness and, second, austere, unambiguous writing. The most common flaws in Help screens, then, are functional confusion (attempting to serve several functions at once) and murky, prosaic messages.

An effective Help screen anticipates the user; it knows what caused the impasse, and what information will release the jam. Not only is it context-sensitive, but it is also alert to the kind of information most appropriate.

In other words, the Help screen not only knows the prompt or field or transaction that caused the user to press the hot button, it also knows whether the user needs **instruction** (procedures, directives) or **reference** (tables of

Exhibit 14.3.4a: Prosey Help Panel

```
CONVERSION - Page 35

Calendar Creator Plus 4.0 should be installed in a separate directory. Copy any
.CAL files from older versions into that directory. Calendar Creator Plus will
automatically convert these old version files to the new version. When the
program detects old version files upon initial startup, changing default event
list path, or running Copy/Merge, it will ask if you want to continue the
conversion process. Press F10 to continue, or Esc to continue without
converting the old files.

At the Text Style prompt, use +/- to select the default text style to be
applied to all events in ALL event lists converted. You can always use the F2
Change Style option on the View Year screen to change the text style of all
events in a file to a new text style. The event list name will be shortened
from 28 characters to 27 characters but otherwise the new file will be exactly
the same.




PgDn Next Page    PgUp Prev Page    Home Table of Contents   End Index   ESC Return
```

allowable values, definitions of cryptic terms, menus of options). Well-made Help screens perform *one* of these two functions. Usually, they do not provide general teaching or orientation (because that's rarely what the user wants); they do not mix instruction with reference.

What most often causes the failure of a Help screen is the attempt to do everything in one panel. Exhibit 14.3.4a, for example, crammed as it is with potentially useful information, will probably frustrate the person asking for help.

If the developers have limited the support plan to one-Help-screen-per-context, then a central problem is deciding which *single, precise function* the screen should perform:

- **Long procedure**—rare and usually inappropriate
- **Short procedure**—in a few terse statements, checklist style, how to complete the transaction causing the impasse

- **Full table of permitted values**—an extended table, sometimes needing more than one panel, with an alphabetical or numerical list of terms, definitions, options, and so forth
- **Short table**—a quick reference with a short menu of allowable entries or responses to the field or prompt

To repeat, when users ask for Help, they need *one* of these. If the needed form of support is lacking, or if it is embedded in a screenful of collateral information, the Help screen may fail.

The current feeling among human factors psychologists and screen designers is that Help screens should contain few, if any, paragraphs. That is, unless there is a compelling reason, the information in a Help screen should be austere, factual. Even simple procedures can be rendered unusable by the paragraph format.

Exhibit 14.3.4b: Paragraph Help vs. Procedural Help

Before

```
To copy objects:
Select the object(s) to be copied, then
use Rearrange Copy. When the box appears
around the object(s), move the box to
the new location. Switch pages if neces-
sary. Press <ENTER>.
```

After

```
To copy objects:
1. Select the object(s) to be copied.
2. Choose Rearrange Copy.
3. When box surrounds the object(s),
   move the box to new location.
4. (Switch pages if necessary.)
5. Press <ENTER>.
```

14.3.5 Designing a '90s Help Screen

Today's users, many of whom work at PCs or "workstations," expect much more from Help screens than their mainframe- and terminal-using predecessors. In the '90s, Help must be context-sensitive; it should not obscure the application when invoked; and it should interact with the main application.

In earlier days of computing, even before most people worked at a video display, a Help facility was little more than a long file (book) stored adjacent to the program it supported. Typically, users at an impasse would ask for help and receive a table of contents (the first menu), from which they would choose pages to read. Afterward, they would return to their application, armed with the missing knowledge, and resume working. In theory.

This traditional notion of Help, which is still everywhere, flies in the face of the relevant human factors. Often, users don't know what they need to know; often, if they are lucky enough to find it, they forget en route to the original screen that created the problem.

Today the demands are higher. A Help screen should be context- or field-sensitive. That is, the system should know the likely problem that caused the impasse. (If the system guesses wrong, then the Help screen should provide a mechanism to explore other possibilities.) When calling for Help produces a menu, it should be short and relevant.

Moreover, invoking Help should not blank out or hide the original screen. With today's graphics adapters and operating systems, it should be possible for the Help panel to appear in a part of the screen that does not overlap the original field or prompt. (It is no longer acceptable to have the original screen disappear and a new one replace it.)

Finally, the Help screen should be more than a document. When it contains a list of allowable values or options—as many do—users should be able to *select* them from the Help panel itself. In other words, it should interact with the application.

Exhibit 14.3.5a: Field Tied to Help Screen

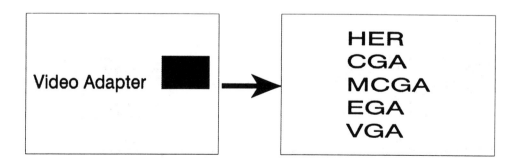

Exhibit 14.3.5b: Field Tied to Short Help Menu

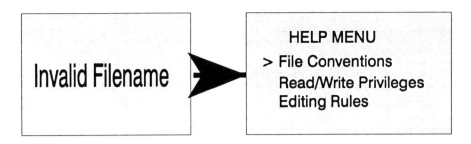

Exhibit 14.3.5c: Popup Help Screen with Interaction

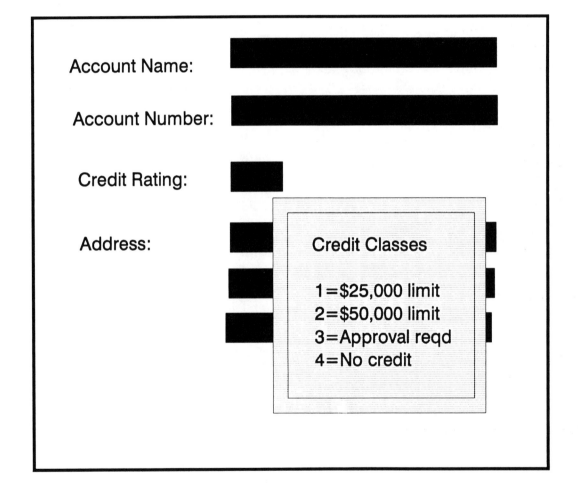

14.4 Computer-Based Training: Two Approaches

Computer-based training is a form of instruction in which either the presentation of materials, the sequence of instruction, or the management of the students is handled by a computer. Sometimes called "tutorials," such programs appear in two main forms: linear and interactive.

Generally, a Help screen is the wrong place to offer orientation to neophyte users. Rather, the appropriate form of online support for the brand new user is the tutorial, or computer-based training (CBT) program.

CBT is also known as computer-based education (CBE) or computer-aided instruction (CAI), and by a few other names. Moreover, various users of the terms may have rather different notions of their meaning. Generally, CBT is a form of teaching in which some or all of the following are handled by (mediated by) a computer:

- **Presentation**—Part or all of the instruction appears on the computer display. (This was not always so; in the early days of CAI the pictorial part of lessons was often in workbooks or on microfiche readers.)

- **Sequence**—The order in which the teaching materials and associated questions are presented is determined by a computer program. This program is either a unique linear sequence or, in more sophisticated examples, a program that branches according to the user's responses. Some programs also remember the user's last session and start the instruction from an appropriate point.

- **Management**—CBT can also manage the progress of students. It can assess amount and quality of learning and feed this information either to the student or to the

student's superiors. It can document (prove) the training received by students, thereby satisfying quality assurance or security demands.

CBT programs fit into two broad categories. **Linear** programs have a fixed sequence. All students move through the lessons in the same order; their only options are to move forward or to quit. In contrast, **branching** programs follow the paths of the users' interests, or of their competence. Users either choose what they want to learn or the computer, which periodically tests the users, chooses paths that fit their instructional deficits.

The software used to develop CBT programs is usually called authoring software. (Nowadays, it is also possible to create training programs with Hypertext and Demo software.) The skills needed to develop effective CBT are those of an instructional designer (or educational media specialist). The average programmer or technical writer is ill-prepared—and lacks the necessary patience—to build a frame-by-frame curriculum. As with orientation support in general, the pace of instruction in CBT is usually so slow that it makes the average writer uncomfortable. ("What!" exclaims the writer. "A whole screen just to explain F keys!")

A further problem is the extensive testing that even a simple CBT program must have. As arduous as it is to test manuals, testing tutorials is harder and more expensive. If a program branches, for example, one should really test all of the branches. (And if there are too many to test, we need a testing specialist to select a reliable sample of paths.)

Even though new software has made CBT easier to implement, it is still expensive—especially when it is tested thoroughly. When, then, is it a good documentation choice?

CBT replaces classroom orientation for neophyte users. (It is less effective, of course, because it does not deal as well with the anxieties of the new user.) It is most economical when

- there are large numbers of such users
- there is so much employee turnover that it becomes impractical to schedule training sessions
- the content of the orientation material does not need to be changed frequently

CBT, then, is a somewhat inferior alternative to stand-up training. It is, however, a superior alternative to most paper publications.

Organizations cannot expect large classes of clerical and subprofessional employees to learn the rudiments of systems from books. The user earlier called Reader X (a person who lacks the skills to learn from complicated books and, as a result, has lost the confidence to try) much prefers CBT to paper publications. Indeed, CBT lets people previously believed to be poor readers reveal themselves to be just as literate as their Reader Y associates. In this way CBT has the potential to liberate large groups of employees from the limited opportunities imposed on them by their discomfort with books.

14.4.1 CBT in a Straight Line

Computer-based training in the linear mode is a one-path program. To move forward, users must get the right answers. If they fail, they may either try again or give up (quit).

The simplest form of linear CBT is usually called a demo, a kind of slide presentation in which the users do little more than tap a key to say when they are ready for the next slide. (Some demos do not offer even that option; they are programmed to show a series of slides at predetermined intervals. Such programs are more likely to be used for presentations at meetings than for training.)

By asking for the next screen, users in effect report that they have learned the content of the current screen and are ready for more. (Some developers call these training programs "tours" or "guided tours.") Exhibit 14.4.1a is an example.

Exhibit 14.4.1a: Demo with Options to Go Ahead or Quit

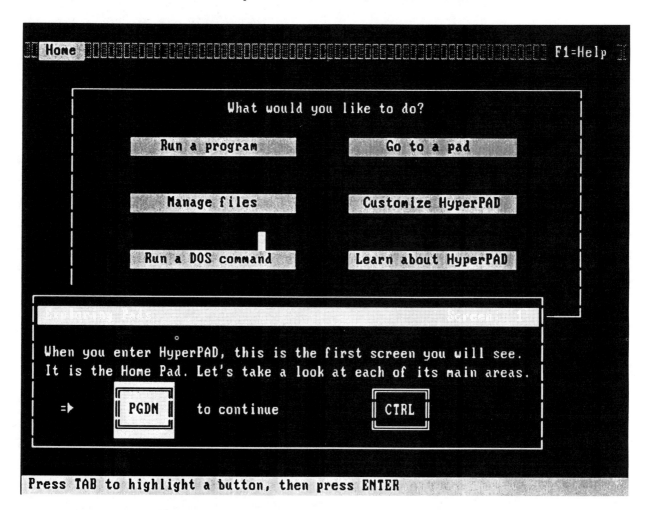

Although it is possible to provide orientation with a slide show, most instructional designers believe that there must be some more active test of the user's understanding. In a linear training program, the user must get the right answer to move ahead. Consider the example in Exhibit 14.4.1b. The options are few: Provide the right answer or else. (Note how the program gives users three chances and keeps track of their progress.)

Of course, the flaw in one-path training programs is that, eventually, the curriculum moves to the next topic, even though the user might not have ever guessed the right answer. And a reader who does not like to ask for assistance (like Reader X) may get deeper and deeper into the program without learning.

The alternative is a program that usually *will not move forward* until the users *prove* they understand.

Exhibit 14.4.1b: DOS Tutorial with Limited Number of Trials/Answer

```
                              Pathnames
                              ═══════════

                                        Assume the shown subdirectory
    Root    Level 1       Level 2       structure (only directories are
    Dir     Subdir        Subdir        shown, not files).

    A: ─┐                 │ BOOK        This series of subdirectories was
        │ WORDPROC ───────┤             setup to categorize various files
        │                 │ MEMOS       developed by a word processor.
        │
                          │ LETTERS     Let's move around in the structure.
═══════════════════════════════════════════════════════════════════════

A test -- What is the pathname from the root to:

Subdirectory LETTERS                    _____




            Please type an answer...You are on try 1 of 3
```

14.4.2 CBT That Branches and Interacts

In some CBT programs, there are many paths, depending on the user's responses or interests. In classic computer-aided instruction, each multiple choice answer determines a different "next" screen. In modern interactive instruction, the learner drives the sequence of presentation.

Many CBT programs are multi-path, modeled after the "programmed texts" of the '50s and the CAI of the '60s.

As in the programmed text, only one answer or behavior will advance the user to the logically "next" item of instruction. But, unlike the single path program, multi-path programs contain special frames of instruction keyed to the precise error of the student. In Exhibits 14.4.2b and 14.4.2c, for example, Option 2 (<F2> Choose Quit...) is the correct answer. But if the user selects Option 3 (<F3> Turn off...), there is a remedial panel aimed at that particular confusion of terms.

Exhibit 14.4.2a: Flow Logic for Branching CBT

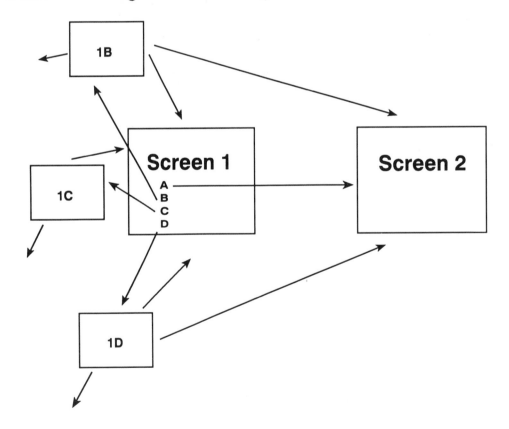

Even this complex branching program, though, may seem simple in comparison with today's interactive training programs. Nowadays it is not uncommon for the CBT program to be linked to a fully operational word processor or spreadsheet, so that the users' responses can include a full range of data transactions, not just selections from a multiple choice menu.

Also, with improvements in computer video, CBT can now be linked to video and sound materials, so that the users' actions can drive the sequence of scenes in an instructional movie.

CBT, in fact, is converging on the most sophisticated form of computer-aided instruction: flight simulation training. Eventually, with the right video and sufficient memory and speed, users in training will be able to "fly" systems in a simulation mode.

Exhibit 14.4.2b: Multiple Choice Screen

To end a session, you should

<F1> Type "End" on the prompt line

<F2> Choose QUIT from the File Menu

<F3> Turn off the power to the computer

<F4> ....

<F5> ....

Exhibit 14.4.2c: Remedial Screen

SORRY. If you turn off the power while working on a file, you will lose the contents of that file.

To end a session, you should

<F1> Type "End" on the prompt line

<F2> Choose QUIT from the File Menu

<F3> Turn off the power to the computer

<F4> ....

<F5> ....

14.5 Improved Support through Better User Interfaces

Documentation, even online documentation, is external to the system or product. The best support, however, is internal: interfaces that eliminate the need not only for manuals but Help screens as well. The easiest improvement is better menus. More ambitious improvements involve what is called WIMP: Windows, Icons, Mouse, Pointer.

The traditional role of the documentor is to support, enhance, supplement. In some cases, the manual writer's job is to compensate for, or ameliorate, the flaws in an underdesigned, undertested system. Writers of manuals can nearly always see ways to improve the system; writers of Help screens can nearly always see ways to reduce the need for the Help screen.

Inevitably, documentors are tempted to improve the design of the system itself, especially the user interface. Unlike many programmers, who consider the user interface or "front end" an afterthought, documentors see it as essential to the *reliability* of the system.

The two main fronts for attacking the problem are, first, reducing the *memory burden* on the users and, second, reducing the number of *keystrokes per transaction*.

The attack on memory burden consists in substituting recognition for recall, so that users do not have to remember command syntax or program and file names.

The screen in Exhibit 14.5a illustrates the ease with which the most often used DOS trans-

Exhibit 14.5a: A DOS Shell

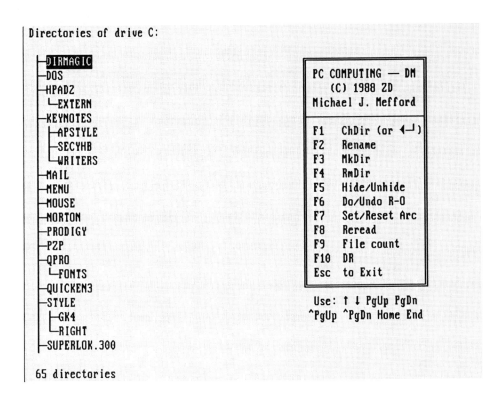

actions can be activated by selecting from a menu.

Not only do such "shell" programs spare the user the need to remember command syntax, they also reduce the need to type. Depending on the action, the user can start a process either by pressing a single F-key or by moving a cursor bar and pressing the <Enter> key (or by clicking a mouse).

While these innovations seem ordinary and unremarkable to today's users, they are in fact an important benefit of the PC revolution of the '80s. (Note that users of large computers learned these techniques from developers of PC software.) Indeed, nearly every innovation in user interfaces has come from the PC industry, only to be adopted later by the allegedly more sophisticated computer systems.

The PC's most visible contribution is the graphical user interface (GUI), in which symbols and pictures (icons) are substituted for words. One argument supporting GUIs is that these graphic entities are "intuitive"—at best understandable without training, at worst easy to remember. Indeed, graphic interfaces are usually discussed as part of a constellation of related techniques:

- **Windows**—the conceptualization of data processing as a scheme of screens or panels that can be overlapped, tiled, moved, sized, and otherwise manipulated by users as they move from file to file or task to task
- **Icons**—pictures that represent actions and entities in the system, so that by selecting the right picture one activates the desired process
- **Mouse**—one of many devices that can move a cursor bar or selection arrow to any of the fields or icons or coordinates of the screen
- **Pointer**—an arrow, hand, or other symbol that marks the user's selection on the screen

Taken together, this cluster of ideas (called WIMP and usually attributed to the inventive genius of Alan Kay [see "Computer Software," *Scientific American*, September 1984]) has become a de facto standard for interface design in the '90s.

Exhibit 14.5b: Graphical User Interface (GUI)

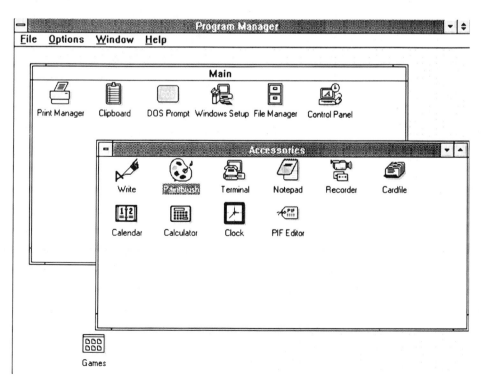

14.5.1 Writing Better Menus

Effective menus are written clearly in the user's vocabulary. They do not limit themselves to single-word entries, and they distinguish between application options *and* navigation options. *In sophisticated systems, menus need "secondary menus" to elaborate or explain their meaning.*

The most cost-effective way to enhance the usability of a system is to convert freeform procedures and command syntax to menu-driven options. (That was the main finding of the early 1980s.) The next most cost-effective project is to *rewrite all the menus to make them more intelligible.*

And the most straightforward task of all is to examine the language of the menus themselves. What appears on a computer screen should be clear, business language, wherever possible in the vocabulary of the users. What should *not* appear is programming terminology (unless, of course, the application is for programmers).

One does not write *execute* for *choose* or *abort* for *withdraw request.* (And one does not start numbered lists with 0 either.) Indeed, consider the absurdity of the most commonly seen menu:

Abort, Retry, Ignore, Fail

Not only is it a monstrous insensitivity to include a word like "abort" on a business display, it is also just plain bad writing. What does each of these options mean? What will happen if I choose Ignore, for example? (How many long-term DOS users have no idea?)

Menus should be free from words that irritate. And they should also be free from words that have different meanings in the users' vocabularies, like *default.*

There are all sorts of myths about menus. Some writers think that menus should contain one-word options. Unfortunately, limiting menu choices to one word makes it nearly impossible to communicate the complicated meanings of these choices.

For example, what's the difference between *copy* and *replicate* on a certain widely used graphics menu? (Answer: *copy* makes one copy of a shape; *replicate* makes as many copies as the user wants.) Not only is *replicate* a poor name for the option (*reproduce* or *duplicate* would have been clearer), the very idea of making this distinction in one-word labels is absurd. What is wrong with *one copy* and *many copies*?

How many pull-down menus give us the choice between *open* and *import.* How many new users know the difference? And even among the experienced, how many know intuitively which files need to be imported?

The one-word myth is a cousin of the "magic number" myth. The world is filled with technical writers who have read George Miller's classic paper on memory, "The Magic Number Seven, Plus or Minus Two" (*Psychological Review*, 63 (2), 1956). On the basis of this reading (or more likely a 200-word summary), they conclude that all lists on screens should be limited to seven items. Thus, they turn one menu into three or four. But no such conclusion can be derived from Miller's paper, or anyone else's either.

Menus should also distinguish between *application options* and *navigation options.* In the early days of menus, such choices as "Exit the System" were included on the menu of options. The newer convention is to separate the *navigational* choices (exiting, moving forward or backward, invoking Help) to a separate zone of the screen, where they can be activated by F-keys or "buttons" selected with a mouse.

Finally, with complicated options and alternatives, the best plan may be a **secondary menu**:

an additional phrase, line of text, or further menu that clarifies what no writer could pack into a well-chosen word or two.

Exhibit 14.5.1a: Menu Separating Application Options from Navigation

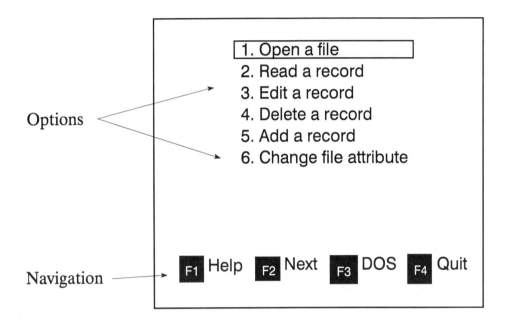

Exhibit 14.5.1b: Secondary Menu with Explanation

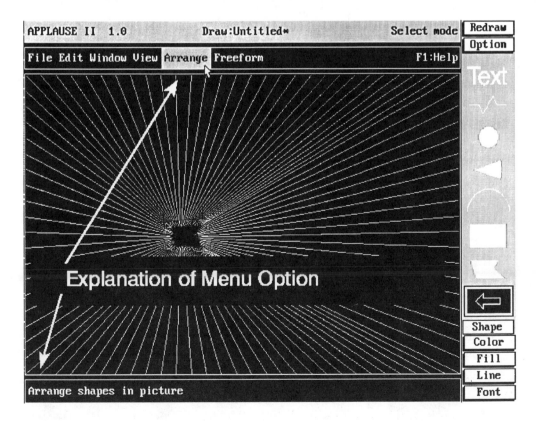

14.5.2 Reducing the Memory Burden: Windows and Icons

The less users must memorize, the better. Nearly every improvement in interface design reduces the memory burden for the user. By substituting windows and icons—or other graphic devices—for command syntax, systems reduce the number of memory-related errors and, thereby, the need for manuals and Help screens.

The **window** is a key element in the contemporary approach to interface design. Its advocates claim that it is a more realistic analogy to the way people actually think and work than the traditional computer screen.

In a window-based system, all the operating software and applications on a computer are assigned a hierarchical stack of panels, screens, or windows. Running a program becomes "bringing up a window for that program"; examining a file becomes "opening a window" for that file. Furthermore, because users often want to look at several files at once, or run one program while not shutting off another, windows technology

Exhibit 14.5.2a: Multiple Windows

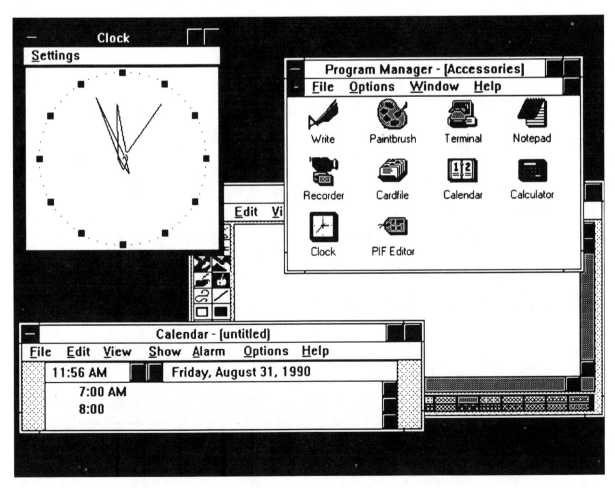

allows several windows or panels to be visible at once. Moreover, the users may move and size the windows to fit them workably on a single screen.

Windows technology eliminates a large memory burden for many users. Once they learn to manipulate the windows themselves, there is no longer any need to write commands that change directories or start programs. In many cases, there is no longer a need to convert file formats. Most routine data transactions—copying, moving, saving, listing, printing—are handled with the same relatively simple point-and-shoot manipulations. And the more multi-tasking (running several applications at once and exchanging data across them) one does, the greater the advantage.

Used frequently in conjunction with windows are **icons**. Icons are more than representations: they *do* what they depict. An icon of a trash can, for example, deletes files, while an icon of a slide projector runs a screen demo.

Again, the intuitiveness of icons—the notion that users will know what they are without being told—is supposed to reduce the memory burden. With little or no orientation users should recog-

nize the icons for application programs and be able to distinguish them from the icons for files.

But there are a few problems with this notion. (I, for one, would be unable to interpret most icons if they did not have text labels.) Even on bit-mapped screens, many icons communicate no clear picture at all. The "tool" icons on most painting and publishing programs mean nothing to users until they read about what the icons do. Moreover, the sequencing and rules for the use of these icons are not intuitive.

Icons are often obscure to the average user. Imagine their value to visually impaired users, who generally prefer interfaces that depend on large letter characters, touch typing, or speech synthesis. (*Note*: Systems that help the visually impaired need ASCII text; usually they cannot interpret bit-mapped documents.)

The combination of well-drawn icons with word labels is a powerful memory aid. But the "intuitiveness" of most icons is something of a myth. And there are many applications in which the system overhead needed to support the WIMP interface is hard to defend.

Exhibit 14.5.2b: Are Icons Intuitive?

14.5.3 Reducing Keystrokes: Mouse and Pointers

Usually, the fewer the keystrokes, the fewer the errors, the fewer the error messages, the less the need for external support. With few exceptions, pointing beats typing, so that the use of the mouse/pointer combination can eliminate much of the need for external support.

In the early 1980s, much of the computer-based office technology was designed for touch typists. Astonishingly, early menus were often labeled with letters of the alphabet, so that a user might be expected to find and press the letter *C*, for example.

Today's user is rarely a touch typist. In fact, a good proportion of the people who spend their days typing into a computer have no typing skills at all. It follows, then, that the longer the string of characters to be typed correctly, the greater the chance of an error. (The error rate seems to rise exponentially with the length of the string.)

Again, menus eliminate much of the problem—especially when they are augmented with pointers or "cursor bars." Most users find it

Exhibit 14.5.3: Pull-down Menu with Alternate Accelerator Keys

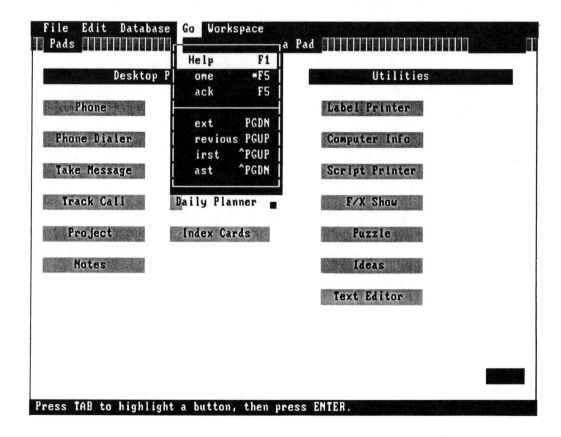

easier to point-and-shoot than to type even one character. (*Note*: A large minority of users prefer "accelerator keys" to the cursor bar; they believe them to be faster and no more error-prone. Moreover, there is a case to be made against mouse-driven pointers: namely, that users may overshoot the mark, often clicking on an unintended choice.)

The mouse/pointer interface carries menu logic still further. In effect, every raster-point on the screen becomes a selectable option. One can put the "cursor" anywhere, and mouse movement is very nearly intuitive (after a few minutes of experimentation).

The mouse is one of a class of devices that move the pointer continuously over a fine grid, or, alternatively, across all the fields on the screen. Tracballs, light pens, and joysticks transform a series of keystrokes (even if they are only tabs and spacebars) into a rapid, fluid, movement of the hand. (There are also helmet-mounted cursor aiming devices, foot-pedal cursors, and even marvelous inventions that allow paralyzed people to move a cursor by puffing on a tube.)

The obvious application for such analog motions is drawing, but this is hardly the most popular application. Smoothly moving pointers can switch quickly among choosing an item from a traditional menu, "dragging" a window into a new position, or freehand sketching with the mouse as stylus. To enable this variety of uses, most systems have more than one pointer to indicate the *mode* of the mouse. Just as most word processors change the cursor to differentiate "insert" mode from "overstrike" mode, most WIMP systems mix arrows, hands, blinking cursors, and a variety of drawing/painting implements to indicate what the mouse is doing.

14.6 Improved Support through Hypertext

Hypertext is a form of communication in which messages are stored at the nodes of a network; readers move from node to node according to their interests (rather than in a fixed sequence imposed by the author). Hypertext is emerging as a preferred form of internal support for resourceful and scholarly users.

The usability of a paper document is largely determined by how much branching, skipping, and detouring is asked of the reader; the more overhead (the effort needed to assemble the right sequence of words and pictures), the less usable the document. But because using technical texts nearly always entails a good bit of this branching, one could argue that being skilled with books means, simply, being able to use easily books that would ordinarily be considered unusable.

Hypertext is mainly a form of reading in which, thanks to computer technology, the reader experiences none of this overhead. That is, in a hypertext document, the reader exerts no more effort in jumping to a page far away than in reading the next page. (In effect, all the pages in a hypertext document are equally close and equally easy to reach.)

The more visionary advocates for hypertext (for example, Ted Nelson in *Literary Machines*) extend its reach to include not only the various pages of a single document, but all the pages in any or all documents. From this point of view, anyone with a computer or terminal can be linked with all the world's information stores and navigate through them with only the slightest exertion.

Hypertext as a notion has been with us for as long as electronic computers. Hypertext as a fact of user support has been with us for as long as Apple has been bundling HyperCard software with its Macintosh products. Since this development, most hypertext software has been built on a simple but powerful model. The material in the hyperdocument is organized into modules or chunks, known variously as cards, pages, panels, or pads. When users view one of these entities on the screen, the screen contains both buttons (icons that activate certain navigational moves) and highlighted terms (words or phrases in an alternate color or font). Users may move to the next screen either by selecting one of the buttons or by selecting one of the highlighted terms.

To repeat, hypertext is a *way of reading*. Instead of the linear organization inherent to books, the organization of the hypertext document is a kind of network in which, to put it simply, there is sometimes no predetermined "next page." Unless programmed otherwise, any node (page) can be linked to any other, so that the effort of moving between them is identical, and so that the ability to retrace one's steps is assured.

Hypertext is not a form of Help, nor even a form of user documentation (although it can be adapted to both purposes). It is, in effect, an alternative to traditional user interfaces and support methods. It is the emerging preference of resourceful and scholarly users, a model for research in the twenty-first century. In conjunction with high-volume storage media, like CD-ROM, it is a natural and exciting improvement on the traditional unabridged dictionary or encyclopedia; and it is increasingly beloved of Bible and Shakespeare students.

But where does hypertext fit into the problem of user documentation? The answers are not yet clear. Any large document that is put online—

such as the library of publications for a computer family—will be far more usable and accessible if conceived of as a hypertext network. With hypertext, for example, the technical assistance personnel who answer phones for the major hardware and software companies can search their company's entire technical library in pursuit of answers to users' questions.

Hypertext is also a programming utility that can be used to attach Help screens, or other support, to existing programs. For hypertext refers not only to the documents but to the usually simple programming languages that we use to create hypertext products. With such programs, documenters can invent and install their own Help screens, even for applications purchased from a third party.

Exhibit 14.6: Hypertext Screen

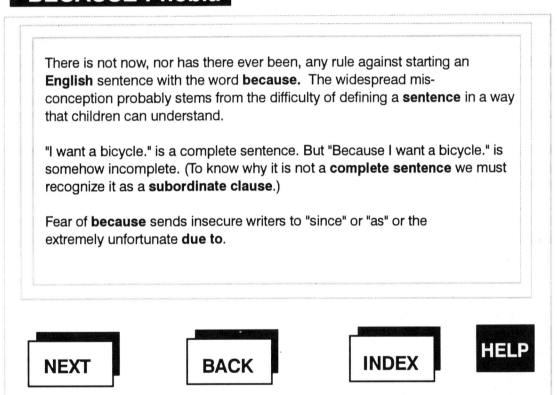

BECAUSE-Phobia

There is not now, nor has there ever been, any rule against starting an **English** sentence with the word **because**. The widespread misconception probably stems from the difficulty of defining a **sentence** in a way that children can understand.

"I want a bicycle." is a complete sentence. But "Because I want a bicycle." is somehow incomplete. (To know why it is not a **complete sentence** we must recognize it as a **subordinate clause**.)

Fear of **because** sends insecure writers to "since" or "as" or the extremely unfortunate **due to**.

NEXT   BACK   INDEX   HELP

14.6.1 Using Hypertext as Help

With a hypertext programming tool, it becomes easy to attach Help screens to particular fields or elements in the application. In more ambitious uses, ordinary Help screens are supplemented with the ability to let users explore any other file that might be helpful—including material from other systems and applications.

Hypertext is both a kind of online document *and* a tool for creating such documents. In its latter sense, as a desktop programming resource, hypertext products can be used to create Help screens and then tie them to specific application screens. This may seem unremarkable, merely substituting a Help button for the now traditional <F1>.

But the difference is less in the result than in the method. Using a hypertext development tool, *the writer can create and implement the Help screens without much assistance from the programmer.* This may help overcome the constraints imposed on user support technologists by programmers, who are sometimes impatient with the creation of Help utilities. In other words, writers willing to invest the week of work needed to learn one of the hypertext scripting languages are free to create the best Help utility they can invent and then attach it (also with hypertext) to the application. This method can even be used to add Help to purchased software products.

Beyond this important change of method, though, there are also changes in the notion of Help itself. Using the hypertext conventions, users can link a Help screen to any of the highlighted terms on the application screen. Moreover, the Help screens themselves may contain highlighted fields and buttons, so that the user can explore a topic or theme until it is exhausted.

A warning: Most users and operators of computers, communication products, and other programmable technology, do *not* really want elaborate "hyperhelp." What most people need when they press <F1> is a quick reference solution to the problem that has stopped their progress. Ordinarily, they neither want nor need long lessons filled with fascinating tributaries of information.

Unfortunately, the problem of the simple Help screen—field-sensitive Help for a routine business screen—has been long solved by professional writers. Writing the Help screens for an order-entry system, for example, is only slightly less boring to an experienced professional writer than writing copy for a parts catalog. As a result, writers tend to be more and more attracted to the exotic possibilities of hypertext approaches to Help. Presentations on hypertext dominate today's meetings of technical writers the way desktop publishing dominated a few years ago.

To put it a bit too simply: Hypertext is for users with curiosity and sense of adventure about the material on the screen. These users are only a minority of the people in need of user support.

Exhibit 14.6.1: Flow for Hyperhelp

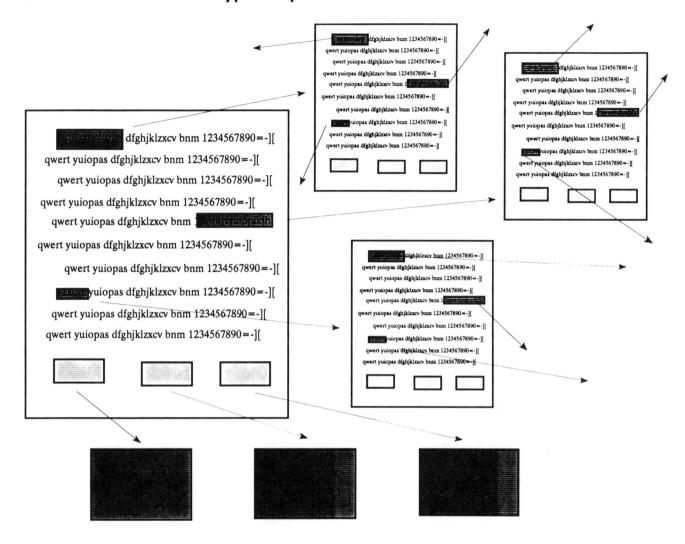

14.6.2 Using Hypermedia as Help

Usually, the material at each node of a hypertext network is, as you would expect, text. *When something other than text is used—graphics, animation, video, sound—we call the technology hypermedia.*

Hypertext programs can be written for nearly any computer with any level of sophistication. Some use text characters exclusively and can run, therefore, on the earliest generations of computers or the "dumbest" of terminals. Most, however, are written for computers with bit-mapped screens—HyperCard for the Macintosh being the best-known example—and make inventive use of graphics.

The most apparent incorporation of graphics into hypertext is the use of icons (pictorial representations of processes). But more interesting is the use of graphics as the materials themselves so that, for example, "clicking on" the name of a city brings up not only a page full of text about that city but a set of maps.

The most interesting hypertext demonstrations at gatherings of technical writers usually involve high-resolution pictures to supplement courses in anatomy or physiology. These products generally use video technology that is a generation or two more advanced than what is currently affordable in business. (Indeed, surveys suggest that the speed with which screens "refresh" themselves—a measure of sheer computing power—is the best predictor of participant interest.)

When something other than text is at the nodes, the network is often called hypermedia. Obviously, pictures may be either still or animated, using computer animation. Thus, certain forms of Help can be provided in moving pictures, such as demonstrations of physical processes (like installing a device).

Certain computer systems, moreover, have video technology that allows conventional analog video to be shown on the same monitor. So a user asking for Help could even see a high-resolution training video.

Hypermedia can also make use of audio—either analog or digital. There can be *spoken* Help screens or, in rare cases, specific sounds that might be relevant. Computers that talk will be with us much sooner than computers that listen, and they have tremendous implications not only for those who do not read well but also for thousands of visually impaired persons. (Please remember that WIMP and GUI technology are not friendly to the visually impaired.)

In principle, anything that can be digitized—including things we can scarcely imagine today—can be stored at the node of a hypermedia network and activated with a key or mouse or even a sound. As fanciful as it may seem to predict the linking of all the world's text resources, hypermedia enthusiasts anticipate the eventual linking of *all information resources, in all media.*

Again, this kind of adventuresome thinking goes way beyond the needs of supporting ordinary business or government users in the 1990s. But certain innovations—like audio Help and animated demonstrations—might be feasible in the near term.

Exhibit 14.6.2: Hypermedia Schematic

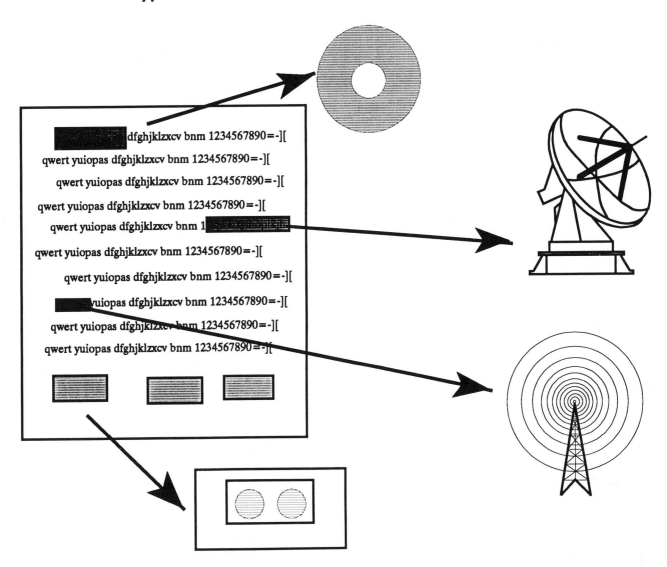

15. AFTERWORD: INTO THE NEXT CENTURY

15.1 Improved Support through the Three *I*'s

In the next century, much of the need for user documentation and other external support will be eliminated through intelligibility *(clear, unambiguous screens and messages),* insurance *(protection from errors and bad paths), and* insight *(applications that forgive individual differences among users).*

We are already well on the road to systems that support themselves without the need for an elaborate support envelope. Indeed, were it not for the software industry's current policy of making every product as feature-rich as possible, we might have already seen the demise of the user manual.

There are three broad strategies for eliminating manuals, Help screens, training programs, and other external information products: *intelligibility, insurance,* and *insight.*

Intelligibility means that screens and messages contain clear, unambiguous, grammatically correct statements written in a vocabulary the user understands. They have been tested for understandability (and have passed!) and they have been converted from their scolding tone to a helpful one. "Invalid number of parameters" becomes, for example, "There is an unneeded space after the drive letter."

There is no more easily attainable and effective improvement than this one. Every menu

Exhibit 15.1: Three *I*'s

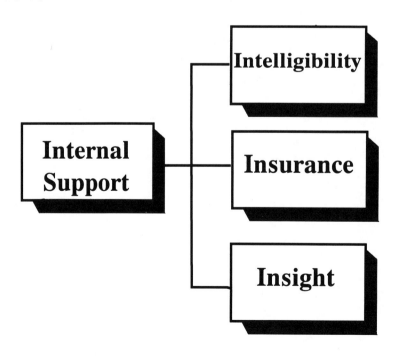

label, every prompt, every "dialog box," every system and error message—everything, including those absurd legal disclaimers and impenetrable "readme" screens—should be written, or at least edited, by someone who knows the difference between a well-made and an ill-made sentence. Moreover, all of them should be tested with representative readers.

A well-designed user interface also gives the user **insurance**: the liberating feeling that trial and error will not lead to serious errors and consequences. The beauty of the pull-down menu, for example, is not just that recognizing replaces remembering and pointing replaces typing. It also *cordons off forbidden and inappropriate moves*: the "fuzzy" options that cannot be selected.

There are milder forms of insurance as well: questions and warnings before you start a routine that will erase data; rehearsal options that allow you to preview the consequences of your choice before you commit to it.

But there is nothing like a closed door, a blocked path, to give readers the expansive assurance that as long as they choose from the active options, they cannot break, erase, or destroy anything. John Carroll describes a project in which, during the learning stage, the system interrupts users before they can complete risky transactions; this temporary measure is called an interface with "training wheels" (*The Nurnberg Funnel*, Chapter 7).

Safe trial and error is, for most users in most applications, a faster way to solve problems than consulting a manual.

Intelligibility and insurance are clearly within the scope of current technology. Only **insight** will demand some programming improvements. An insightful program is one that knows what you mean, even when you have not used the precise names and syntax the system expects. Just as a spelling-checker will usually guess what was meant, insightful software will form a hypothesis about your intention, ask you (Do you mean ? . .), and make the change. Over time, moreover, it will keep track of your individual proclivities and compensate for them without asking.

In short, insightful software will be at least as resourceful as young typists used to be. Surely, that is not asking too much of artificial intelligence.

15.2 Author Power: An Agenda for Documentors

Documentors have all the tools and technology they need to produce outstanding work. What they often lack, however, is will and assertiveness. In the '90s, documentors should insist on certain standards of corporate conduct.

Although technical writers are better trained and equipped than ever, they are too often victimized by old-fashioned notions of documentation and out-dated models for their work. They must assert their rights and insist on proper treatment for themselves and their projects.

Here is an agenda for the '90s:

1. **Campaign for the assignment of competent writers.** Even today, sophisticated firms are still assigning too many writing projects to people without training, skills, or experience. Many companies use the technical writing slot for employees whose "real" jobs have been eliminated. Still others think that manuals can be written by secretaries and clerks.

   Moreover, programmers usually cannot write usable user documentation. And they probably cannot write understandable system messages and Help screens either.

   Professional writers should protest this practice whenever they encounter it; they should make clear to their employers that manual-writing is a mature craft and that it is therefore wasteful to assign inept and unqualified people to the job. The work invariably takes longer (if it is ever finished) and the quality is usually poor.

2. **Do not tolerate substandard writing—on page or screen.** Punctuation, spelling, grammar, usage, idiom, economy of style—all these matter a great deal in user documentation. Resist and challenge anyone who says otherwise. Also, tell everyone that all the requirements for a well-made page are exag-

gerated on the screen, especially the need for "white space." Beware of people who want to conserve screen space; they are more dangerous than those who want to conserve paper.

3. **Insist on written specifications for documents.** Never begin to write a document until there is a written, official description of the scope of the piece. (Spoken, informal understandings are nearly useless.) Assure especially that the audiences are named and that relevant assumptions about their previous knowledge and training are spelled out. When possible, the specification should be complete enough to be the basis for a usability/acceptability test.

4. **Resist unrealistic deadlines.** When you are given a deadline for a project, be sure that it is based on an assessment of the work to be done—not a capricious date determined by other factors, and not pulled out of a hat. Do *not* accept impossible timetables; negotiate, resist, refuse. Distance yourself from any manager who values deadlines ahead of quality.

5. **Insist that all documents be tested.** An untested publication is *full of bugs* at every "level of edit," from misspellings to inconsistencies to confusions of purpose. No deadline justifies the distribution of an error-filled, misleading, sloppy, embarrassing publication. Remember that all most users "see" of systems are the user interface and the user documentation; their first impression should not be disheartening.

6. **Reduce the need for documentation, wherever you can.** Ironically, a central goal for a '90s documentor is to reduce the need for manuals. (Later in this decade, there will be a similar campaign to reduce the need for Help screens.)

Usually, people who write user documentation become experts on what makes systems hard to use. (The harder a procedure is to perform, the harder to write the instructions.) Therefore, instead of patiently documenting everything that comes along, they should attack what they believe are bad systems, bad procedures, and bad screens.

Before explaining a quirky or error-prone task, the documentor should learn why it is that way. And absent a suitable explanation, the documentor should demand an improvement. (In the mid-'90s, desktop programming will enable documentors to fix the problems themselves.)

7. **Apply the Golden Rule of User Documentation.** The Golden Rule for documentors is

> *Do not do unto your readers what you have hated when it was done unto you.*

Because writers use so much diverse software these days, they gain an added insight into user support. Namely, they have been victimized by several poor manuals. Most writers would do better work if they merely remembered their own frustrations with poor manuals and vowed never to inflict similar pain on their readers. Even when their employers—with excuses of time and tradition—seem to be asking them to.

Ultimately, consideration for readers is the central, ethical issue for writers. Good writing, you see, involves *sacrifice*. As a writer, you must exert yourself more so that your readers can exert themselves less. Write clearly, simply, and honestly—because it is the right thing to do.

APPENDIXES

Appendix A:
Excerpt from User Support Plan

The material below is an excerpt from a hypothetical user support plan.

MEMO

TO: Resource Requirements Forecaster Project Team

FROM: Executive Committee

RE: Support Policy

University RRF product is meant for the planning and budgeting officials of a medium- to large-sized university. It will be used by a small cadre of people—usually fewer than 20 individuals, **all of whom are presumed to have experience with spreadsheets running on PCs.**

As a condition of sale, we shall provide a one-day training program to any group of 20 identified by the customer. Most users, therefore, will **receive their orientation from a consultant, rather than a manual.**

In a typical customer university, the RRF will be used 3-5 times a week during a 3-4 month planning/budgeting activity. **During that peak activity, we hope that users will not need to consult their manuals.** We expect, though, that after a long hiatus, users may need their manuals to jog their memories. (Our marketing people are thinking of offering an annual seminar, partly as a way of encouraging the sale of upgrades and add-ons.)

We think that any manuals should be in the large (8.5 x 11) format and that the spine should display the product name prominently.

RRF TOPIC LIST

(Partial)

Hardware requirements for PC-based RRF
Mainframe version of RRF
Installing RRF on PC systems
     Memory Options/Limitations
Getting technical assistance
Technical assistance policies

The nature of resource requirements forecasting
Planning errors associated with "manual" forecasting
Key terms associated with direct expense
Key terms associated with indirect expense
Benefits of the RRF as a stand-alone product
Benefits of the RRF when linked with other forecast-
  ing modules

The nature of forecasting
The differences between forecasting and simulation
How spreadsheets work
How linked spreadsheets work

Setup: Setting up student/teaching staff factors
        Setting up student/other staff factors
        Setting up student/other service factors
        Setting up noninstructional activity factors
        Prorating indirect costs and subventions
        Setting up the State Aid module
        Setting up the tuition module
        Setting up the financial aid module
        Setting up the portfolio module
Editing the setup screens
Creating multiple versions of setup for simulation

Using the opening menu
Using the Enrollment Demand Screen
     Linking with the Enrollment Forecaster
Using the Service Demand Screen
Using the State Aid screen
Using the Portfolio Forecaster Screen

The Standard Reports:Text
The Standard Reports: Graphical
Customized reports for selected programs and periods
Simulations of multiple economic assumptions
Sensitivity checks for variables
"Spec Mode": Anchored (targeted) projections

RRF Audience List

1 University Presidents/Vice Presidents—
  Prospective Clients
1 University Presidents/Vice Presidents—**Actual
  Clients**
1 University I/S Managers
1 Institutional Research Specialists
1 Information Officers
1 Labor Negotiators
1 Mid-level Administrators **without spreadsheet
  experience***
1 RRF Trainers **

* We believe that there is still a small group of
  prospective users without spreadsheet experience—and that they should be supported.

** If we are successful, we'll need to engage
  several new trainers, who will need support as
  well.

RRF User Support Plan

The RRF Project Team has come up with the following plan for information products in support of the University Resource Requirements Forecaster.

Publications

1. *The RRF Brochure*—a 20-30 page marketing piece intended to sell the benefits of RR simulations (glossy, 6x9 format; signature bound)

2. *The RRF Setup and Installation Guide*—a 50-75 page technical manual aimed at the single person on each site responsible for installing the product and entering the university's defaults (loose-leaf, 8.5x11, updated by page supplements; one-per-site)

3. *Using the Resource Requirements Forecaster*—a 30-40 page manual for all those users NOT responsible for establishing or altering the university's factors (but permitting simulation of alternative factors); mainly a series of screens with associated instructions (loose-leaf, 6x9 format, 2 colors [data in a second color]; all exhibit screens filled in with representative data)
   NOTE: 2 & 3 are combined in an *RRF Instructor's Guide*

4. *Spreadsheets: An Introduction*—a 10-12 page brochure for users unfamiliar with spreadsheets, in which all the illustrations are from RRF, usable as an advertisement for our product

5. *RRF Template*—a laminated keyboard template containing the definitions for the shifted and unshifted function keys (12x2) and also other key quick-reference material for the system

Online

1. *2-Tiered Help Facility*—context sensitive Help in which the first screen shows allowable values for the field and the second gives a procedural plan for the whole screen (about 250 screens)

2. *RRF Tutorial Disk*—an orientation demo that runs passively for first-time users and in a branching, hypertext mode for experienced users needing a refresher

Appendix B:
Illustrative Modular Outlines for User Manuals

The following exhibits contain Tables of Contents that grew from modular outlines. That is, each heading corresponds to one two-page module. (In some cases, there are no modules for Chapter or Section names.)

1. Outline for a University Resource Requirements Forecaster

1. How the RRF Works
   1.1 How Activities Consume Direct Resources
   1.2 How Indirect Resources are Attributed to the Direct

2. How to Work the RRF
   2.1 Entering Enrollment Demand
   2.2 Entering Service Demand
   2.3 Stipulating the Scope of the Analysis
   2.4 Generating the Report

3. Using the RRF as a Simulator
   3.1 Setting-Up "What If" Forecasts
   3.2 Prediction Mode: Adjusting the Factors
      3.2.1 Adjusting Staff/Student Ratios
      3.2.2 Adjusting Other Service Ratios
      3.2.3 Adjusting Inflation and Economic Factors
   3.3 Target Mode: Finding Factors that Meet Specs
   3.4 Testing the Sensitivity of the Factors

4. Linking the RRF with Other Modules
   4.1 Linking with the Enrollment Forecaster
   4.2 Linking with the State-Aid Analyzer
   4.3 Linking with the Economic Forecaster

5. Presenting the Results
   5.1 Generating Text/Statistical Reports
   5.2 Generating Graphical Versions of the Forecasts

2. Outline for a Supervisor's Guide to Documentation Center

1. Lines of Authorization in the Doc Center
    1.1 Authority: Who May Approve a Job
    1.2 How to Assign Job Priorities
    1.3 Table of Organization
    1.4 Table of Duties and Responsibilities
    1.5 Eight Preconditions for the Use of WP Facilities

2. How to Configure the Doc Center System
    2.1 Selecting the Software/Application
    2.2 Selecting Printers or Plotters
    2.3 Selecting Scanners

3. Level I Jobs: Basic Correspondence
    3.1 Defining a Document File
    3.2 Entering Text
    3.3 Printing a Review Copy
    3.4 Editing the Text
        3.4.1 The Twenty Most Common First-Draft Errors
        3.4.2 The Three Most Difficult Revisions
    3.5 Printing the Finished Copy
    3.6 Sending the Finished Copy Through the Electronic Mail

4. Level II Jobs: Advanced or Technical Documents
    4.1 Assembling Documents from Older Documents
    4.2 Merging Document Variables
    4.3 Performing Arithmetic within the Software
    4.4 Generating a Mailing/Distribution File
    4.5 Interpreting Ambiguous Input (Default Rules)

5. Policy: Logging and Storing of **All** Documents

6. Policy: Protecting the Confidentiality of Our Clients

7. Policy: Resisting Pressure from Originators and Managers

3. Outline for a Manual for Creating a Special Purpose Phone Network (Excerpt)

Chapter 3. Building and Verifying a TERRITORY-MAP

3.1 Each TERRITORY-SET is a Telephone Database

3.2 How to Start a New TERRITORY-SET

3.3 How to Enter the Data for One Sector's TERRITORY-SET

3.4 How to Transfer Set Data from the AZ-60 to the AZ-190

3.5 How to Create an Auxiliary Database Called INTERIM

3.6 How to LOAD Data from INTERIM to TERRITORY-SET

3.7 How to Get Access to GRAF-MAP

3.8 Five Requirements for All Maps

3.10 How to "Introduce" a Map to the System

3.11 The Most Efficient Way to Enter Map Data

3.12 How to Use the ZOOM Feature of GRAF-MAP

3.13 How to Manipulate a Map

  .1 Adding Buildings and Nodes

  .2 Changing Buildings and Nodes

  .3 Adding Terminal Data

  .4 Changing a Terminal Profile

  .5 Modifying the Terminal/Site Matrix

  .6 Adding a Segment Between Two Nodes

3.14 How to Validate Entered Data

3.15 Review: The TERRITORY-SET Checklist

4. Outline for a Guide for the E-POST Electronic Mailing System

1. Conventions Used in This Manual

2. Three Reasons to Use E-POST Instead of Ordinary Mail

3. Knowing When You have Received an E-POST Message

4. Getting Access to E-POST

5. Getting Help from E-POST

6. Using the E-POST: An Overview

7. Using the E-POST Index
   7.1 Getting Access to the Index
   7.2 Searching for U-Names by Last Name
   7.3 Searching for U-Names by Account Number or ID
   7.4 Locating U-Names on Other OGR Computers

8. Using Distribution Lists: Overview
   8.1 Setting Up a Distribution List
   8.2 Changing a Distribution List
   8.3 Directing E-POST Messages through a Distribution List

9. Printing E-POST Messages: Three Methods
   9.1 Printing the Screen
   9.2 Printing with the STORE Facility
       9.2.1 Saving an E-POST Message as a STORE
       9.2.2 Printing the STORE File
   9.3 Printing with the Word Processing Facility
       9.3.1 Saving the E-POST Message as a Document
       9.3.2 Sending the Document to SCRIBE-15
       9.3.3 Printing the Message by its SCRIBE-15 Document ID

10. How to Send an Existing File as an E-Post Message

11. How to File E-POST Messages in Secondary Indexes

12. E-POST Commands: A Glossary

Appendix C:
Illustrative Module Specs

The exhibits below show various examples of a module specifications.

HEADLINE: Importing Text Saves Time

SUMMARY: Although it is possible to enter text directly into the PageDesign layout screen, it is usually easier and faster to import it from one of the supported word processors or from an ASCII text file. This way you can use existing documents, scanned text, or downloaded files without needing to retype or re-enter data.

EXHIBITS:

DIRECT ENTRY   **IMPORTED**

- Headings   ▪ WP, WS, Word, PFS files
- Editorial adjustments   ▪ ASCII (text) files
                                     ▪ PCQ, TRX, GOF, CCC graphics
                                     ▪ Network downloads
                                     ▪ Scanned input

NOTES: Stress the number of popular word proceesors supported. Do not explain the procedure, but emphasize how much time can be saved and how much flexibility is provided.

HEADLINE: Converting an "Unrecognized" Graphic Format

SUMMARY: PageDesign recognizes the 5 most common graphics file formats. In addition, it includes a utility, GRAFFIX, that will capture the screens from other graphics formats and convert them to the PageDesign (.PDS) format. To use this feature, load GRAFFIX as a resident program; display the graphic from the other program; hit <Alt>-G; follow the instructions on the screen.

EXHIBITS:

PD Utility Menu   Graphic Screen   GRAFFIX Screen

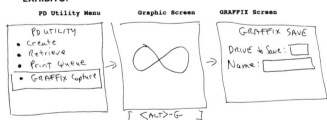

NOTES: Mention compatibility problems with Windows. Suggest Windows Clipboard as alternative process.

HEADLINE: IMPORT: Incorporating Another File

SUMMARY: *Importing* is the process of incorporating a text or
                             graphic file into your page layout, without typing in
                             the text or drawing the graphic. By choosing *Import*
                             from the file menu, files made with programs other
                             than Page Design can be included in your layout.

EXHIBITS:

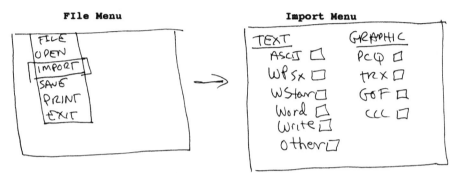

NOTES: Mention that non-PageDesign files cannot be reached by
                             choosing the "open" option from the file menu.

HEADLINE: Appendix D: Importable Graphics Files

SUMMARY: The table below lists the graphics files that can be
                             imported directly into Page Design and the popular
                             programs that use those formats.

EXHIBITS:

TABLE: Listing importable graphics formats--

  - PCQ   (Hotpaints, Draw This, Yale-Draw....)
  - TRX   (Quality-Draw, CheapCad, Sketchpad...)
  - GOF   (Pixelpot, PrintSlave, ClipWorld...)
  - CCC   (VectorLine, Tracer, Fractal Fun...)

NOTES: Mention GRAFFIX for unsupported formats.

Appendix D:
Illustrative 2-Page Modules

NOTE: The eight examples here were written by my clients and students. They are reproduced as they appeared, showing a variety of styles and printing technologies.

1.

---

1. INTRODUCTION TO THIS GUIDE   1.4   HOW TO USE THIS GUIDE
                                 1.4.2 EXPLANATION OF FORMAT USED
                                       IN TEXT AND DISPLAY

This guide has been formatted on the basis of a two-page spread in which written text has been positioned on the left and examples or displays positioned on the right.

Throughout the text, stars (*), BOLDING, SMALL CAPITAL LETTERS, and *italics* identify specific data elements as follows:

- the asterisk (*) has been used in sample screen displays to indicate cursor position.

- *italics* has been used to show data which has been keyed into the input fields[1] by the user.

- bolding with LARGE CAPITALS has been used to indicate that the text is refering to a specific key on the keyboard.

- SMALL CAPITAL LETTERS has been used to indicate the actual field name used in the screen display.

See Figure 1.4.2 which shows the conventions used in the text to differentiate between data being input by the user, names of keys, and screen prompts.

1. Some of the displays you use will allow you to enter information in certain areas of the display. These areas are called input fields.

<u>Figure 1.4.2</u>
Content of Text: meaning of small caps, bolding, italics, asterisks.

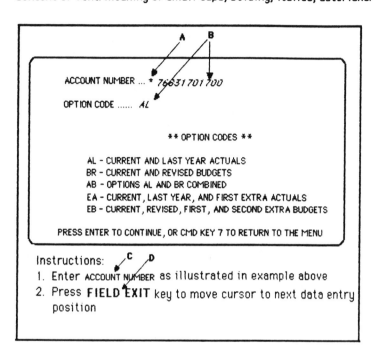

A: Asterisk (*) shows cursor position.
B: *italics* indicates data which has been input or keyed in by user.
C: SMALL CAPITAL LETTERS used to indicate field name shown on screen.
D: **BOLDING WITH LARGE CAPITAL LETTERS** used to refer to a key on the keyboard.

2.

---

18. Three Ways to Store TABLES Commands

You may choose one of three ways to store TABLES commands.

A TABLE MEMBER is a set of TABLES commands stored as a member of a partitioned TABLE FILE. You may store only DEFINE, IDENTIFY LIST, and ADD TRANSFORMATION commands in a TABLE MEMBER.

A BINARY TABLE is a set of TABLES commands similar to a TABLE MEMBER, but encoded in a compact, binary format for quicker and more efficient storage and retrieval. A BINARY TABLE may contain only DEFINE, IDENTIFY LIST, and ADD TRANSFORMATION commands. You cannot alter or edit a BINARY TABLE.

A COMMANDS DATA SET is a set of TABLES commands, stored as card-images, but not as part of a TABLE FILE. A COMMANDS DATA SET may contain any TABLES commands. You may edit or alter a COMMANDS DATA SET.

To create a Table Member or a Commands Data Set, use a text editing program such as UNI-COLL's QED or IBM's EDIT. To create a Table File in which to store Table Members, use BUILD TABLE FILE. To create a Binary Table, use WRITE BINARY TABLE.

To recall a Table Member, use READ TABLE, after you have SELECTed the Table File to which the member belongs with SELECT TABLE FILE. To recall a Commands Data Set, use READ COMMANDS DATA SET. To recall a Binary Table, use READ BINARY TABLE.

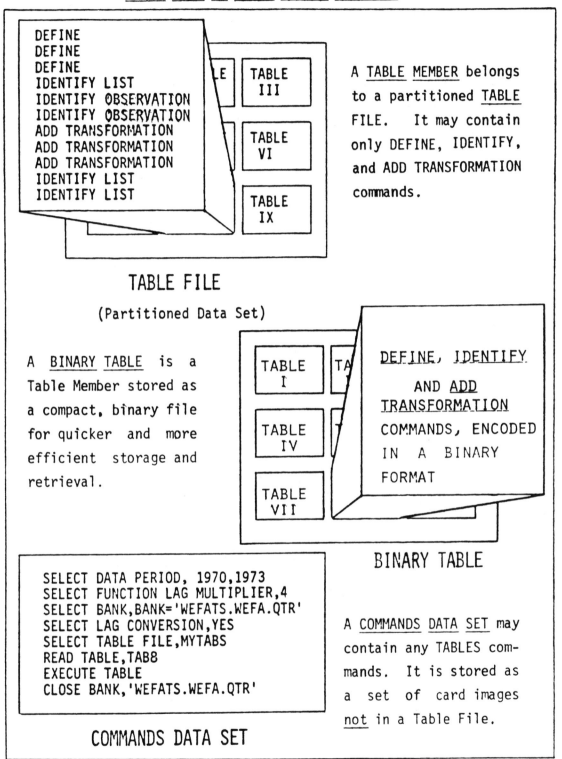

```
DEFINE
DEFINE
DEFINE
IDENTIFY LIST
IDENTIFY OBSERVATION
IDENTIFY OBSERVATION
ADD TRANSFORMATION
ADD TRANSFORMATION
ADD TRANSFORMATION
IDENTIFY LIST
IDENTIFY LIST
```

TABLE
III

TABLE
VI

TABLE
IX

A TABLE MEMBER belongs to a partitioned TABLE FILE. It may contain only DEFINE, IDENTIFY, and ADD TRANSFORMATION commands.

TABLE FILE

(Partitioned Data Set)

A BINARY TABLE is a Table Member stored as a compact, binary file for quicker and more efficient storage and retrieval.

TABLE
I

TABLE
IV

TABLE
VII

DEFINE, IDENTIFY AND ADD TRANSFORMATION COMMANDS, ENCODED IN A BINARY FORMAT

BINARY TABLE

```
SELECT DATA PERIOD, 1970,1973
SELECT FUNCTION LAG MULTIPLIER,4
SELECT BANK,BANK='WEFATS.WEFA.QTR'
SELECT LAG CONVERSION,YES
SELECT TABLE FILE,MYTABS
READ TABLE,TAB8
EXECUTE TABLE
CLOSE BANK,'WEFATS.WEFA.QTR'
```

A COMMANDS DATA SET may contain any TABLES commands. It is stored as a set of card images not in a Table File.

COMMANDS DATA SET

2.2   Vector Command – Drawing Straight Lines

```
===================================================================
The Vector command draws lines and dots.  There are four basic
options available to specify how the vector command can be used.
===================================================================
```

The Vector command draws straight lines from the current
cursor position to a specified end point. This end point
can be specified as an absolute or relative position.

With the Pixel Vector (PV) system and the Write Command
Multiplier option, you can draw a line in a specified
direction for a specified distance.

Drawing a Dot

If no co-ordinates are supplied, the V command draws a dot
at the current cursor position:

V[ ]

Drawing a Straight Line

If you supply coordinates, the V command draws a straight
line from the current location to a specified position. The
general form of this command is the same as for the position
command. The format is

V[x,y]

You can use absolute, relative, or mixed addresses, as in
the Position command. For example, if you wish to draw a
line to an X value of 200, and 100 pixels lower than the
current Y value, use this command:

V[200,+100]

Drawing by Direction Using the Pixel Vector (PV) Option

You can also use the offset directions, in the format

Vn

For example,

V4

Once again, you must repeat the direction many times to produce a visible line, or use a multiplier value. A multiplier value can be set either as a temporary write control, or as an option of the Write Control command. It sets the number of pixels written in a given direction. For example, a multiplier of 10 produces a vector 10 pixels long for a single command. For example:

W(M10)V4

Drawing a Closed Figure

You can use the Vector command for drawing closed pictures. V(B) establishes your initial position and (E) returns you to that position. So if you use V(B) and then specify some vectors, (E) closes the figure. For example,

V(B) [+60][,+60][-100](E)

draws a rectangle.

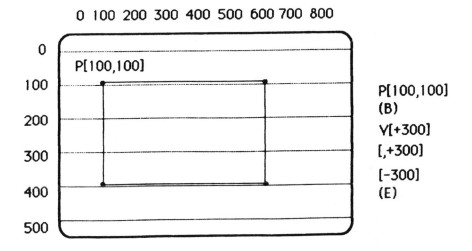

0 100 200 300 400 500 600 700 800

P[100,100]

P[100,100]
(B)
V[+300]
[,+300]
[-300]
(E)

6.2.2 Opening Configuration Views from Another View or from a Command Response Window

You can open a configuration view from any other view by selecting the icon and then choosing Open Configuration View... from the Functions menu. If the selected icon represents more than one component, you select a specific component from the list in the SELECT NETWORK COMPONENT dialog. You can open a configuration view from a command response window by selecting the text line for a particular component and then choosing Open Configuration View... from the Functions menu.

From Another View

With a network, component, or configuration view on screen, you can open a component view by identifying the component you want to be the subject of the view:

1. Select a component icon. (To select a link, click on the circle at the link midpoint.)

   Note: Do not select more than one component. Configuration views can only be opened one at a time.

2. Choose Open Configuration View... from the Functions menu. (Or press <Ctrl-F3>.)

   If the selected icon represents only one component, the view will appear.

If the icon represents more than one component, the SELECT NETWORK COMPONENTS dialog appears so that you can select a single component from the group. Each component represented by the icon is listed in the dialog's scrollable box. Each row in the list corresponds to one component, showing the network ID, domain ID, component ID, and status.

To select a component from the SELECT NETWORK COMPONENT dialog:

   Double-click on the line that identifies the component you want as the subject of your view.

   You can also select the line and then click OK. (Or press <Enter>.)

The dialog closes and the configuration view appears.

From a Command Response Window

When you issue a command from the Display, Changes, or Test menu, Net Manager displays a command response window. These windows display text lines identifying the command's subject component and the result of the command. These text lines can be selected just as icons can in views.

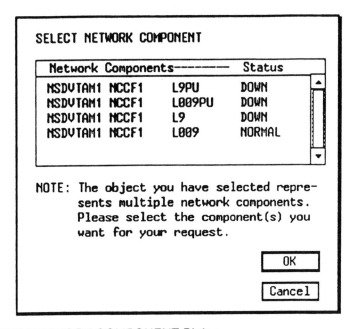

Figure 6.5 The SELECT NETWORK COMPONENT Dialog

G11NM030

```
┌──┬─────────────────────────────────────────────────────────┬──┐
│ ⋈│          Display Status Command Response                 │ ◆│
├──┼─────────────────────────────────────────────────────────┼──┤
│ ? │NSDVTAM1   NCCF1    NSINCPB       04/06/88      14:04:59 │ ▲│
│HELP│ID        Type          Status       Information        │  │
│   │                                                         │  │
│   │ NSINCPB   FEP/HOST      Normal                          │  │
│   │  LoadDump PROCESS       Down                            │  │
│   │ 030-L     CHANNEL       Unavailable  I/O operations = 10083│ │
│   │ L000      LINK          Down         Never has been activated│ │
│   │ L011      LINK          NotConfig    Control ownership has changed│ │
│   │ L005      LINK          NotConfig    Control ownership has changed│ │
│   │ L006      LINK          Normal                          │  │
│   │▐L009      LINK          Normal       Activated by operator▌│ │
│   │ L010      LINK          Down         Never has been activated│ │
│   │ L007      LINK          Normal                          │ ▼│
│   │ L008      LINK          Normal                          │  │
│ ◄ │                                                    ► │ ▮│
└──┴─────────────────────────────────────────────────────────┴──┘
```

Figure 6.6 Selecting a Text Line to Display a Configuration View

G11NM051

To open a configuration view from a command response window:

1. Select the line containing the component.

   Note: Do not select more than one component.

2. Choose Open Configuration View... from the Functions menu. (Or press <Ctrl-F3>.)

   The configuration view appears.

5.

XP
MASTER CROSS REFERENCE LIST BY PART NUMBER, MFG CODE

PURPOSE:

The XP **TRANSACTION** is designed for use by:

Inventory Management personnel

Cataloging personnel

Any Kuwait Air Force/Air Defence personnel interested in looking for a NIIN to go with a Part Number and Mfr. Code.

The XP **TRANSACTION** should be used to:

Find a NIIN for a Part Number and Mfr. Code.

DATA BASE USED:

MCRDB — Master Cross Reference Data Base

TRANSACTION MODES ALLOWED:

6 — **RETRIEVE**

TRANSACTION KEY REQUIRED:

Part Number --(32 **CHARACTERS**)

Manufacturer's Code ---(5 **CHARACTERS**)

NOTE:

If the Part Number/Mfr. Code that you enter has more than one NIIN,

You will get a **SECONDARY KEY SELECTION SCREEN** that lists all of the NIINs that have the Part Number/Mfr. Code you entered.

ERROR MESSAGES:

1000 — NO PART NUMBERS FOUND FOR NIIN ENTERED

1001 — NIIN ENTERED NOT FOUND

1002 — PART NUMBER/MFG. CODE ENTERED NOT FOUND — USE XP TRANSACTION TO CHECK

1003 — PART NUMBER/MFG. CODE AND/OR NIIN MUST BE ENTERED

1004 — MULTIPLE NIINS EXIST FOR PART NUMBER/MFG. CODE ENTERED — USE XP TRANSACTION TO DETERMINE NIIN DESIRED — RE-ENTER TRANSACTION WITH NIIN

1005 — UNKNOWN PART NUMBER CAN NOT BE ENTERED

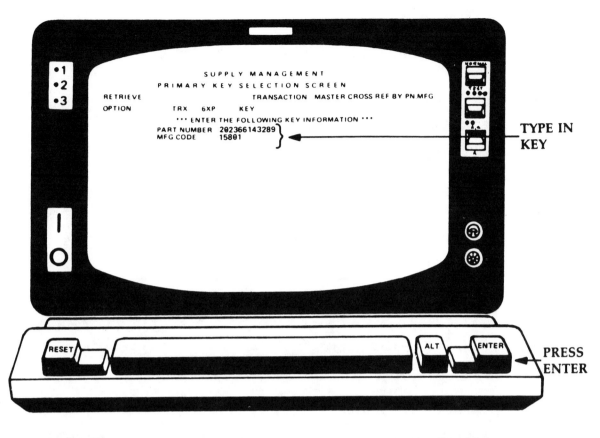

TYPE IN
KEY

PRESS
ENTER

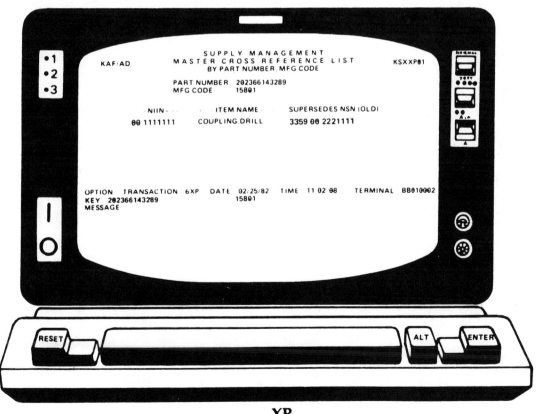

XP
SCREENS

6.

1.5   What the connectors on the back of the RBERT are

> The RBERT installed in the top of middle rack 'C' has several connectors on its back. The picture here explains each. The only power connection (E) is to a standard 120VAC supply. The other connectors are for data.

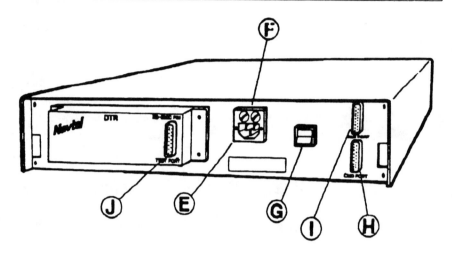

Exhibit 1.5   Back panel of RBERT.

issued: 12/15/88

E	**AC Line Cord Connector:** The line cord supplied with Datatest Remote is a standard three conductor appliance–type AC connector. The ground lead on the cord must be connected to the power ground at the power outlet.
F	**AC Fuse Receptacle:** This holds two 2 AMP, 250 volt, slow blow fuses. Ensure the line cord is removed from the AC supply before changing these fuses.
G	**Power ON/OFF Switch:** This rocker switch turns the AC power to Datatest Remote ON or OFF.
H	**Command Port:** This is a DB 25 female connector that emulates a DTE. Any ASCII terminal connected to this port lets you control Datatest Remote locally. If you connect this port to a dataset, you can control the RBERT through DataPac.
I	**AUX Port:** This is a DB 25 female connector that emulates a DCE. The Aux port connects to RAP A1. *Step-by-step* instructions for doing this follow.
J	**Test Port:** This is an RS–232C interface. This port is the interface which receives the test data.

issued: 12/15/88

2.6 Manually Selecting Pallet Destinations in Standalone Mode

When you are working in Standalone mode, you must use the Standalone screen to select the next workstation for a pallet.

When you are operating in automatic mode, the MHSC automatically chooses the pallet's next destination when you press **[F6]**. If you are operating in manual mode, you must choose the destination.

In standalone mode, the destination screen is displayed when you press the **LOAD RELEASE [F6]** key on the operations screen. The destination screen is shown in Figure 2.

To choose the destination in manual mode:

1. Release the load by pressing **[F6]** on the operations screen. A destination screen will be displayed.

2. Select the destination from the choices displayed across the bottom of the screen: **[F1] - [F8]**.

3. Confirm the choice by pressing **[F21]**, **SEND STANDALONE MOVE REQUEST**. The pallet will be sent on its way and the operations screen will be displayed again. When the pallet clears Lift Table #2, both lift tables are automatically raised to the level of the punch nibbler table.

4. As before, manually move the empty pallet on Lift Table #1 to Lift Table #2.

You have now finished one complete loading and unloading cycle and are ready for the next aluminum sheet.

Figure 1. When you press **[F6]** to release a load in standalone mode, the destination screen shown in Figure 2 is displayed.

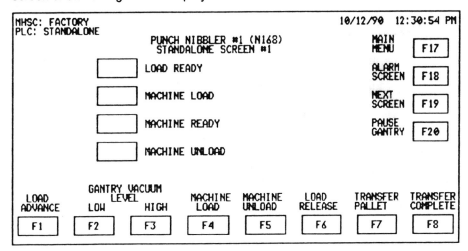

Figure 2. The destination screen lets you choose where to send a pallet. When you press a destination key, the pallet is sent and you are returned to the operations screen.

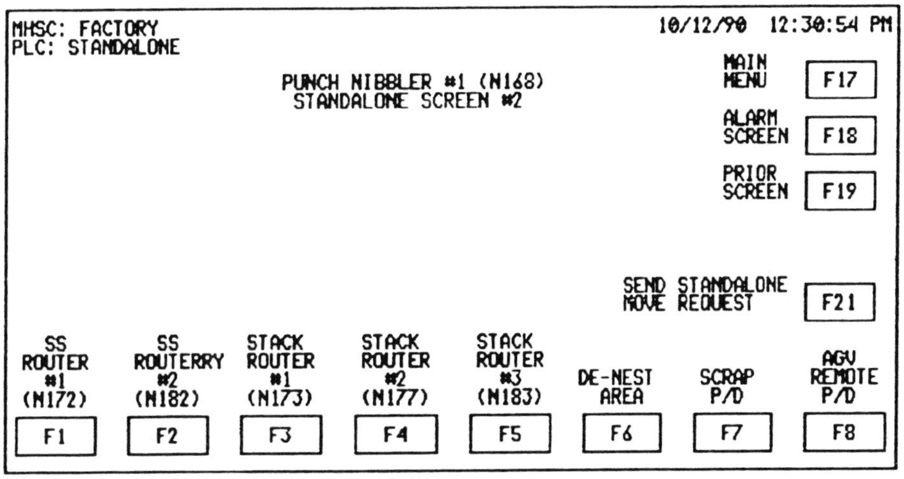

2.2 FORMAT DETAILS FOR THE CREATE PROGRAM INPUT FILE

The INPUT file to CREATE consists of a sequence of individual curves. Each curve is specified by three ordered sections: name and attributes section, independent parameter section, and dependent parameter section.

Curves are specified on the INPUT file one right after another. The curve name card for a new curve begins right after the last card defining the previous curve. An end-of-record (i.e., 7/8/9 card) causes CREATE to stop reading input. A maximum of 250 curves may be specified.

Each curve is specified by the following three ordered sections:

Curve Name and Attributes Section

The following information is specified on these cards:

(1) Curve Name.

(2) Format flags for reading the tables of independent and dependent values. Since these formats are specified for each curve, they may vary for individual curves.

(3) Comments to store on the curve file. These comments are optional and are printed by the AUDIT and PRINT programs.

(4) Number of independent and dependent parameters.

(5) Interpolation and extrapolation codes.

Independent Parameters Section

These cards are used to specify the independent parameter names, units, and values.

Dependent Parameter Section

These cards are used to specify the dependent parameter names, units, and values.

The format details for each of these three curve format sections are discussed in the next three sections of this document. Two examples immediately follow the details and you should refer to them when needed to help your understanding of the formats.

12

INPUT FILE
TO CREATE

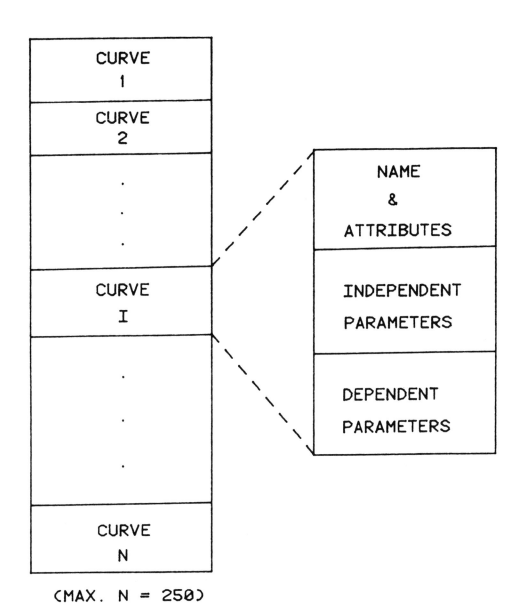

(MAX. N = 250)

13

Appendix E:
Glossary of Selected Terms Used in This Book

The glossary below defines terms invented for this book, or terms that are used here with special or stipulative meanings.

Term	Definition
accessibility	the ease with which information can be found or extracted
artistic stereotype	a method of writing in which most of the effort is in the draft and relatively little in analysis and design
audience	a group of readers with common interests (similar tasks) and common background
availability	the presence or absence of a document
engineer stereotype	a method of writing in which most of the effort is in analysis and design and relatively little in the first draft
GOTO	an unconditional branch; in manuals, entering or exiting a module in the middle
guidance module	part of a manual that teaches an entire process to an experienced reader
headline	a thematic or substantive heading, associated with one module of documentation
help screen	a panel of information that helps users get through an error or impasse OR an access port to a larger file of online technical information
hypertext	an approach to reading, in which information is stored in a network of nodes, which the reader may reach through many paths
information support plan	a plan defining all the user manuals and other information products or services associated with a system
maintainability	the ease with which systems or manuals can be debugged, repaired, and modified
model	representation of a system or product, used to facilitate testing
modular outline	a list of headlines naming each module in a planned manual
module	small, functional, independent entity in a system or document
module specification	a sketch defining the contents—text and exhibits—for a given module
motivational module	a part of a manual that gets readers to perform a task they are reluctant to perform
online documentation	any method in which procedural or reference information is delivered through the computer's display, rather than in paper documents

orientation module	part of a manual that teaches one new task or idea to a neophyte reader
readability	the ease with which a passage can be read, often expressed in grade level of difficulty
redundancy	deliberate repetition and duplication, meant to reduce the burden for the reader and offset the effects of noise and distraction
reference module	a part of manual that serves as auxiliary memory for the user to "look up"
reliability	absence of interruptions and failures
storyboard	a working display showing the specs for each module in a manual, a model for the emerging document
strategic errors	failure to develop the right mix of information products and services
structural errors	failure to organize the contents of a document into the most usable sequence
structured	of a process, developed through top-down analysis and modeling; of a product, organized into modules and the links that couple them
structured documentation	the method whereby principles of structured analysis and design are applied to the writing of publications, especially computer documentation
suitability	the degree to which a manual fits the interests and supports the tasks of the user
tactical errors	failure to edit drafts for clarity and readability
task-oriented	of documentation, defined so as to support users in precisely what they do
usability	the ease with which a system, product, or manual can be used
usability index	the more often the intended reader must skip, branch, or detour, the less usable the book
user documentation	all the information products devised to help users adapt to their computers
user support technology	the profession of assembling information goods and services in a way that contributes to productive, reliable work by users
user:task matrix	an array of topics to be documented and users/readers to be supported; used in defining the mix of documents

Appendix F:
Books and Periodicals for Documentors

*Since about 1982, there has been a steady stream of books about user documentation—manuals and online. The titles below are a **short** list of works that all documentors should know.*

Books About User Documentation

Brockmann, R. John *Writing Better Computer User Documentation* Version 2 (Wiley 1990)

> This single book contains the most exhaustive review of recent research on user documentation, as well as the most comprehensive bibliography.

Carroll, John M. *The Nurnberg Funnel* (MIT Press, 1990)

> A provocative work that lays out the theory of "minimalist" instruction and documentation, certain to be one of the main topics in the 90s.

James, Geoffrey *Document Data Bases* (Van Nostrand Reinhold, 1985)

> A much-discussed work that shows how the largest and most advanced computer systems can be harnessed to the tasks of maintaining and distributing documentation.

Sandra Pakin & Associates *Documentation Development Methodology* (Prentice-Hall, 1984)

> Just the thing to get a novice writer started. And, indeed, companies without publication standards can even adopt the contents of this book as a Standards and Procedures manual for the organization.

Simpson, Henry, and Steven Casey *Developing Effective User Documentation* (McGraw Hill 1988)

> An excellent review of the PC-based tools for managing and producing user documentation.

There are also three exceptionally enlightening anthologies, all of them published by MIT Press:

Barret, Edward (ed) *Text, Context, and Hypertext: Writing with and for the Computer* 1988

Barret, Edward (ed) *The Society of Text: Hypertext, Hypermedia, and the Social Construction of Information* 1989

Doheny-Farina, Stephen (ed) *Effective Documentation: What We Have Learned from Research* 1988

Books About Online Documentation

For anyone interested in online documentation, the essential work is

Horton, William *Designing and Writing Online Documentation* (Wiley 1990)

Also strongly recommended are

Galitz, William *Handbook of Screen Format Design* 3/e (QED Info Sciences 1989)

Rubenstein, R and Hersh *The Human Factor: Designing Computer Systems for People* (Digital Press 1984)

Shneiderman, Ben *Designing the Human Interface* (Addison-Wesley) 1987

For those especially interested in hypertext, I recommend

Shneiderman, Ben, and Greg Kearsley *Hypertext Hands-On* (Addison-Wesley 1989)

And, of course, no study of hypertext could be complete without the incomparable *Literary Machines*, self-published by its author, Ted Nelson.

Books About Clear Writing

Many of the people assigned to work on user documentation are new to the ranks of technical and professional writing. The titles below are among the best general works on how to write clearly, especially about technical and scientific subjects.

Brogan, John *Clear Technical Writing* (McGraw-Hill, 1973)

Strunk, W., and White, E.B. *The Elements of Style (3/e)* (Macmillan, 1979)

Tichy, H. J. *Effective Writing for Engineers, Managers, Scientists* 2/e Wiley 1988

Weiss, Edmond H. *One Hundred Writing Remedies* (Oryx Press, 1990)

Periodicals for Documentors

Articles about user documentation may turn up anywhere, but these are the most reliable places:

Journal of Documentation Project Management, published by Pakin & Associates in Chicago

* *The Journal of Computer Documentation*, Special Interest Group on Documention of the Association for Computing Machinery (ACM SIGDOC)

Technical Communication, the Journal of the Society for Technical Communication, Washington D.C.

IEEE Transactions on Professional Communication

I also recommend the annual conference proceedings of STC, ACM-SIGDOC, and IEEE-PC. These anthologies, available through the respective professional societies, are among the most provocative and useful books on technical communication in general and user documentation in particular. And not the least of their virtues is that most of the papers are written by professional writers. Reading so many pages of clear, well-edited technical writing is often inspiring.

Index

by Linda Webster

User documentation *(continued)*
    testing for, 54-55, 160-65
    two ways to write, 40-41
    "universal task architecture," for, 64-65
    usability of, 7
    usability versus economy in, 32-33
    users' needs and, 5
    work breakdown for, 56-57
User interfaces. *See also* Menus
    DOS shell, 210-11
    icons, 211, 215
    improvements for, 210-17
    improvements in, 188
    mouse, 211, 217
    pointer, 211, 217
    WIMP, 210-11
    windows, 211, 214-15
    writing better menus, 211, 212-13

User support, 180-81, 226-28
User support plan, excerpt from, 232-35
User support planning team, 60-61
User-friendliness, 185
Users
    analysis of, 68-69
    definition of, 4
    needs of, 5

Vogue words, 142, 146

WIMP, 210-11
Windows, 211, 214-15
Within-modules redundancy, 126
Word bugs, 142-43, 146
Work breakdown for user documentation, 56-57
Writing, books on, 260-61

EDMOND H. WEISS, Ph.D.

Edmond H. Weiss, Ph.D., is an independent consultant and lecturer on technical writing, management communication, and documentation. He is also the author of *The Writing System for Engineers and Scientists* (Prentice-Hall, 1982) and *100 Writing Remedies: Practical Exercises for Technical Writing* (Oryx Press, 1990).

Usually, Weiss is traveling North America teaching seminars and consulting for major corporations. At other times, though, he lives in Cherry Hill, New Jersey with his actress wife and writerly daughter.